SLUDGE
and Its
Ultimate Disposal

Edited by
Jack A. Borchardt
William J. Redman
Gordon E. Jones
Richard T. Sprague

ANN ARBOR SCIENCE
PUBLISHERS INC / THE BUTTERWORTH GROUP

PREFACE

This book is a compilation of papers presented at a seminar entitled "Sludge and Its Ultimate Disposal" held at the University of Michigan on January 30, 31 and February 1, 1980.

This seminar was planned by a joint committee representing the Michigan Section of the American Water Works Association and the Michigan Water Pollution Control Association. The conference was sponsored by these organizations in cooperation with the Michigan Department of Natural Resources, the Michigan Department of Public Health and the University of Michigan College of Engineering, Division of Sanitary Engineering.

The course was designed to serve superintendents, municipal officials, plant operators, consulting engineers, technical staffs, manufacturers' representatives and others who have an interest in the subject.

The purpose of the seminar was to explore the subject of sludge and its ultimate disposal in the fields of water and wastewater. This area of interest is changing rapidly as the theory and technology of its subject matter become better understood. The chemical aspects, toxicological problems and administrative difficulties involved in sludge disposal were discussed, along with the problems in the development of adequate standards to control human hazards. The ultimate objective of this exchange of ideas was to provide a reasonable and economical solution to the disposal of the final concentrated pollutants being generated in the water and wastewater industries.

The organizers of the seminar and editors of this volume sincerely appreciate the efforts of all authors who were enthusiastic contributors to the successful seminar and prompt in submitting the papers that make up this book.

Planning Committee

Stuart Bogue
Jack A. Borchardt
James K. Cleland
Liberato D. D'Addona
Frederick T. Eyer

Gordon E. Jones
William J. Redman
Richard T. Sprague
William Wheeler

Jack A. Borchardt is Professor of Sanitary Engineering at the University of Michigan. Prior experience has been with the Michigan Board of Health, the University of Wisconsin, the Wisconsin Board of Health and the U.S. Public Health Service. He received his MS in civil engineering from Carnegie Tech and his PhD in sanitary engineering from the University of Wisconsin. Dr. Borchardt is a registered Professional Engineer in Michigan and Ohio and has served as consultant to the U.S. Navy on pollution. He has received six awards for research and has a number of publications to his credit.

William J. Redman, Training Officer for the Michigan Department of Public Health, received his BS in chemistry from the University of Detroit. He had a long career with the Detroit Metro Water Department during which time he received the Edward Dunbar Rich service award and the Raymond J. Faust Award for outstanding service to the water works industry. He is a member of the American Water Works Association, the Engineering Society of Detroit, and the National Environmental Training Association. He has authored papers dealing with various types of filter operation research.

Richard T. Sprague is State Specialist, On-Land Waste Treatment for the Michigan Department of Natural Resources, Water Quality Division. He received his BS in applied sciences from the University of Michigan and his MS in soil physical chemistry with a minor in sanitary engineering from Michigan State University. He is a member of the Agronomy Society of America, the Soil Science Society of America, the Water Pollution Control Federation and is a Certified Professional Soil Scientist (ARCPACS).

Gordon E. Jones, is a partner in the consulting firm of Wilkins & Wheaton Engineering Company, Kalamazoo, Michigan. He received his BS in civil engineering and his MS in sanitary engineering from the University of Michigan. He is a registered Professional Engineer in Michigan and Indiana, a diplomate of the American Academy of Environmental Engineers and is active in the Water Pollution Control Association and the American Water Works Association. He has authored several papers in the area of wastewater treatment.

CONTENTS

Section V
Regulatory Aspects

SLUDGE: WHAT CAN BE DONE WITH IT?

Joe G. Moore, Jr.

Assistant Administrator
Detroit Water and Sewerage Department
Detroit, Michigan

In January 1979 there were 10,000-20,000 tons of frozen wet sludge piled up on all the unoccupied space of the Detroit Wastewater Treatment Plant. After the spring thaw, the decision was made to remove the sludge. The Detroit Water and Sewerage Department (DWSD) asked for bidders to move the sludge, and received one bid. After enormous effort, all of the sludge was removed.

The Detroit facility still has problems and probably will have them for some years in the future. Sludge will continue to be a major problem because it is increasingly difficult to know what to do with it. The plant treats 600-700 mgd, on an annual average basis, and can quickly generate an enormous quantity of sludge that requires disposal.

Sludge is the unwanted part of the wastewater treatment process—the stepchild, the orphan, of water pollution control and water supply treatment in recent years. As far as water is concerned, sludge is the last thing we have come to in dealing with the environment. No one wants it next door. However, sludge has to be somewhere; it does not and will not disappear. To some degree, we have deluded ourselves in the public media that we have eliminated pollution. However, we have not really stopped water pollution, we have merely moved the pollutants around. There is a very basic physical principle that matter is not destroyed, and we have not destroyed many of the pollutants. We simply have been removing them now for some 15 years in a very concentrated form.

If sludges are burned, their constituents are put back into the air. If they are washed out with scrubbers, we put them back into the water. As we increase the treatment levels for either potable or waste water, we generate increasing quantities of sludge. One of the obvious methods of disposal is to bury it. In many cases, whatever damage it can do is likely to be done again. We have merely chased the pollutants from one medium to another. Fundamentally, we must recognize that we either make pollutants so insoluble that they cannot further influence the water, or we separate them chemically so that the chemical components are more easily dealt with. We are reducing pollutants in the environment; however, the chances are that we are merely preventing their release into the environment rather than removing them after they get there.

The Clean Water Act of 1977 had a rather innocuous amendment. In a cross-reference to a different section, there is a reference to wastewater treatment plant sludge and how the U.S. Environmental Protection Agency (EPA) administrator is supposed to deal with that question. Immediately preceding that, a phrase was substituted for language that had been in the old law, under Subsection c of Section 405: "Each State desiring to administer its own permit program for disposal of sewage sludge subject to Sub-section a. of this Section within its jurisdiction may do so in accordance with Section 402 of this Act." That last phrase, which was substituted for language in the 1972 Act, obscures the significance of that section unless one is familiar with Section 402 of the 1977 act. Section 402 of the original act, continued in the 1977 amendments, prescribes the procedure for the National Pollutant Discharge Elimination System (NPDES) permit procedure, and thus the statutory change means that all future disposal of sludge is subject to NPDES. Sludges are now regarded in the same way as pollutants discharged into the water, and must be permitted in the same way.

Leon Billings, of the Environment Subcommittee of the Senate Committee on Environment and Public Works, has said that "wastes are merely resources out of place...." However, if there is value in waste and sludge, it does not necessarily follow that the value is recoverable in any simple way, process or procedure.

Recycling probably offers the best course for doing something with sludge. The obvious recycling procedure is well known—land application. However, if land application is practiced, either in academic institutions or in wastewater or water supply treatment plants, one should follow it long enough to determine whether substances contained in applied sludge eventually will find their way back into the food chain and thence into the bodies of organisms in the food chain, including the human food chain. Experiments done in Texas record the uptake of various substances into plants that could be used for cattle feed or pasture, to determine whether substances enter the land food chain and perhaps bioaccumulate in land animals in the same way that they bioaccumulate in aquatic organisms.

A similar activity, land reclamation (e.g., the reclamation of strip-mined lands), is an obvious way to use sludge. This use can also encounter interesting difficulties. The city of Detroit, through the DWSD, has been in extensive discussions with Canadian officials and officials of BASF-Wyandotte, a chemical company, about the use of Fire Island, an island in the Detroit River, for the application of Detroit sludge. The island has a levee on the outside. Waste accumulates inside this levee, and water evaporates over a period of time. Nothing grows there because the substances placed there are not conducive to growth. The land might become productive if sludge were added. This possible use of Detroit sludge triggered much publicity and several rather critical statements by officials of the Michigan Department of Natural Resources about the application of sludge on this island, which is under Canadian jurisdiction, and the possibility of disposing Detroit sludge there can no longer be considered. It is conceivable that the use of Detroit sludge on Fire Island might have been of benefit not only to the DWSD, but also to the people of Canada, by restoring that island to some potential use. Again, would you care to have sludge next door to you? In this case, Detroit was told "you cannot put it next door to us."

Land reclamation in remote areas may offer some opportunity for disposal, but it does little good if the cost of transporting sludge to the disposal site is so high that "you can't get there from here." It is possible that we might be able to find some sites in the Detroit metropolitan area for Detroit sludge, again, if sludge can be composted. This requires space, which is at a premium almost anywhere, but certainly more so in a city such as Detroit, with a wastewater treatment plant that generates so much sludge. If sludge can be composted, it might be recycled; again, one must be concerned with substances left in the sludge.

Sludge can be used as a soil conditioner if it is composted. One of the most striking uses of composted sludge is the application of Minneapolis-St. Paul sludge to the mangrove areas of Florida. (This use would also eliminate the mangrove areas of Florida after a while if sludge were applied over the entire area.) A barge of Minneapolis-St. Paul sludge comes down the Mississippi River and through the intracoastal waterway to Tampa-Hillsborough Bay; a long way, which says something about the value of that particular sludge. Of course, it was conditioned adequately before it left. The uptake of substances remaining in the sludge would still be a matter for concern.

Another use of sludge is to generate energy. This use should be of increasing importance, and sludge has been used in various places over the years as an energy source. In Fort Worth, Texas, sludge is digested in anaerobic digesters; the methane is collected and stored, and then used to run the pumps for peak flows of the aeration tanks of the activated sludge process.

In Dallas, where they also use anaerobic digestion, the methane gas is flared. Thus, one finds right next to one another, two different ways of handling the gas. There have been attempts to sell the methane gas from Dallas, but there are some technical problems in making sure that impurities in the methane do not pose a problem in the distribution system.

"Coincineration" is a word that is becoming popular—the burning of sludge with solid waste. One of the interesting problems is often an institutional one. Detroit is well on its way to building an incineration facility for solid waste, but there was never any consideration of mixing Detroit sludge with solid waste. The funding for the solid waste incineration facility comes from the Department of Housing and Urban Development. The project is so far along now that it would possibly delay it to pause at this point and consider the addition of sludge to the solid waste; yet that is a very logical combination. The two departments involved apparently do not cooperate as well as they might, or this could have been worked out when both of them were making their respective plans for future expansion.

Obviously, straight incineration is one way to dispose of sludge. This is the Detroit way, for the time being. However, the current system is energy-intensive; that is, the city buys natural gas to keep the sludge burning.

Obviously, it would be more desirable if one could achieve autogenous burning. That possibility raised the question of dewatering capability of Detroit's equipment. The dewatering capability is inadequate to get the sludge to a concentration of solids that will assure autogenous burning. We may move in that direction in the short run, simply by substituting a different method of dewatering sludge. Also, there are problems; if certain substances are added to the sludge before it is incinerated, incineration is complicated. The energy value of sludge should always be considered.

One of the ways sludge has been used in some places is to combine sludge and solid waste in landfill sites. A gas-collecting system is installed and the methane is collected and used. The Los Angeles County Sanitation Districts have a major landfill site that looks like a gas field, because the gas-collecting pipes stretch across the surface. They sell the methane, after treatment, to a distribution system in a residential area to supplement the natural gas normally supplied in the distribution system. They are examining the possibility of doing this more extensively. The same agency that operates the wastewater treatment facilities also operates the facilities for the disposal of solid wastes; thus, the administrative and institutional problems are overcome with the consolidation of the functions within a single agency.

There will, however, always be some ultimate disposal. Even incineration does not get rid of everything; one ends up with ash. Again, if the pollutants were not changed in the incineration process, they remain in the ash. The question remains: "what do we do with this final residual?"

Anaerobic digestion is a process widely used in other parts of the country. It is one that has been considered by Detroit, but its complications hinder its implementation. They did have some capacity to digest sludge, but they took that system out some years ago. Again, it may be difficult to return to it. If sludge is digested, the gas can be recovered, but there is still a residual. However, digested sludge could be disposed of more easily at landfill sites than the existing Detroit sludge.

There is always the possibility that in the final residual toxic substances remain. The EPA has made it clear that it intends to regulate toxic substances "from the cradle to the grave." There was a major internal dispute that delayed the designation of some toxic substances because there were those outside the water area who insisted that all sludge should be designated as toxic. The effects of that can be realized by identifying the sites where toxic substances can be disposed of in Michigan. There has been such adverse experiences and accompanying publicity with toxic substances in existing disposal sites (e.g., the problem with polychlorinated biphenyls), that it seems very unlikely that if sludge were classified as a toxic substance there would be any place in Michigan where one could put it. The problem is sites: where could you find a place to put toxic substances? Abandoned sites have already received a tremendous amount of publicity. There are sites where substances were discarded that no one thought at the time were toxic, or maybe substances that people knew were toxic and the generators hid them by disposing of them on plant sites. There has been so much publicity that the mere mention that sludge might be a toxic substance will raise difficult disposal problems.

Sludge volume is going to increase. The higher the treatment levels required for water supply and wastewater treatment, the more sludge will be generated. It will all have to go somewhere; it will not vanish into thin air.

There is increasing resistance at the local level to disposal sites; it is extremely difficult to get permits. The city of Detroit has probably spent somewhere in the range of $3-5 million just employing consultants to find sites for Detroit sludge. The volume and level of the problem can be indicated by the Detroit budget. The operating budget for 12 months contains an item of $17 million for the *interim* disposal of sludge. This is in addition to the incineration system onsite. This amount is merely to haul sludge that cannot be burned onsite to disposal sites. That represents a major operation and maintenance expense that will continue indefinitely into the future unless some more permanent solution is found. My recommendation is that Detroit buy its own site and use it exclusively for Detroit sludge. At $15 million per year, the city could buy a sizable portion of the state of Michigan. I think that is the way the city ought to go, since it is confronted with these problems. No one wants to buy the land inside the city of Detroit;

even the city that generates the sludge would prefer that it go next door—or farther, if it could be moved farther.

The state has a responsibility, and I think, in the long run, it is the states that must solve the sludge problem. The state must secure the sites for disposal of the more critical residuals we are likely to generate. I think that states ought to secure and operate toxic substances disposal sites as a matter of state policy. I do not think that two local units of government adjoining one another can solve the problems of disposal sites if there is any possibility that a question of toxicity will be raised. It is difficult for Detroit: mere mention that Detroit sludge might be disposed of in some local community almost precludes any possibility of being able to get through the regulatory process; thus, I think the states have the ultimate responsibility.

Fortunately, in Michigan, the governor has at least gotten the state into the business of trying to find a site for the storage and disposal of radioactive wastes. Whether anyone in the state accepts the fact that this is the way to go, it took a great deal of political courage for the governor to make a recommendation like that. I think the state ought to take the responsibility for disposal sites for critical residuals.

There will be another problem as a consequence of existing statutes. With the industrial waste control programs being installed as a result of the federal Water Pollution Control Act and its 1977 amendments, industries will soon begin to determine ways to handle their wastes onsite, if they can. One of the ways they can handle waste onsite is to remove it and stack it there. There have been some instances of that already in this country, perhaps even in Michigan. That way, they do not have to pay any surcharge because they did it their way on their own site. All that does is to postpone the problem, because it will eventually be found there, or so much will be accumulated that the company begins to lose money. They will ultimately have to do something with it. Then they will have an enormous quantity that will have to be disposed of. The pretreatment requirements will mean that there may be small quantities of critical sludges accumulating all over the state. One of the things we keep telling people about Detroit sludge is that as soon as we get the industrial control program under way and the pretreatment program in place, Detroit sludge will be better than it is. The reply of a Detroit contractor to that was, "yes, but no sludge is better than good sludge."

"Better" sludge does not mean that the pollutants have been eliminated; they have merely been taken out and accumulated elsewhere in smaller quantities. They may be easier to handle in smaller quantities, but if Detroit cannot find a place to put sludge that contains critical substances, industry is not going to find places any more easily. It may be scattered further in the environment, and if it does damage, we may have more difficulty collecting it from a large number of sites than we might have if we put it in one place.

The underground injection control regulations in the Safe Drinking Water Act will have an effect on sludge disposal. One of the classic ways to hide waste some years ago—which may still be going on in some places—is deep well disposal on an industrial site. Just dig a hole and put it under pressure and stick the waste down there somewhere; maybe it will not come up soon or near, and it is hidden—it is gone. Because of oil field activities in Texas, they had dug well disposal regulations and statutes before 1960. A tremendous amount of water in that state has been contaminated by the drilling of oil wells, but, fortunately, have had little contamination from deep well disposal. EPA had an extensive controversy over pits, ponds and lagoons; when is a "well" a well, and when is it a pit, pond or lagoon? All of those things leak, too, and all can pollute ground water. There is an interesting provision of the Safe Drinking Water Act in that particular area. It is the "sole source aquifer provision," and is intended to protect any ground water that is the sole source of a municipal water supply. Its application could close substantial areas of the country to surface development.

When the Resource Conservation and Recovery Act is in full effect, it will also affect sludge. In my view, the sludges should be reduced to the smallest possible volume, put in the best containers that we can devise, and accumulated where we know where it is. Some people have said that the best place to put wastes is in Texas, because there is so much land that is unusable for anything else. That may be true; at least we would know where they are. The way we are going now, we do not know where they are; we do not know what the effect is, and at some future date, they will come back to haunt us, because we have merely moved them around. The obvious place to hide them is to bury them in the ground, and I am afraid that this is what is being done. We are merely obscuring the problem for the long run. Until we solve what we will do with sludge and how we will handle the residuals that end up in sludge, we will not have overcome either water or air pollution. The great risk in the long run is that in achieving adequate water or air pollution controls, we will pollute the land. That could be considerably more difficult to cure than the pollution of water or air.

SECTION I

SLUDGE MANAGEMENT—TECHNICAL ASPECTS

CHAPTER 1

INDUSTRIAL SLUDGE DISPOSAL IN MICHIGAN

John M. Bohunsky, Chief and
Roy E. Schrameck, District Engineer
Water Quality Division, Field Operations Section
Michigan Department of Natural Resources
Lansing, Michigan

Until the 1970s, very little consideration was given to the control of industrial sludge. Such material generally was regarded as a waste product from a manufacturing process to be discarded by the most convenient means possible. Disposal practices were varied and almost always unregulated. Some industries placed sludges in earthen pits behind their manufacturing plants, while still others trucked these materials to landfill sites or to an isolated rural location. Many industries discharged their waste materials into publicly owned treatment works (POTW), a practice that continues to be prevalent in many cities.

Land disposal of sludge was a common method and was considered an acceptable practice in almost all cases. The soils were considered to have an unlimited carrying capacity for sludges of all types. Little was known about the ion exchange and transfer potential of metals to plants or the filtering capacity of the soils. No one suspected that chemicals in the sludge could pass through the soils to any appreciable extent and contaminate the groundwaters.

The first serious consideration given to regulating the disposal of sludges on land followed the passage in 1965 of the Solid Waste Act. This Act was primarily intended to regulate nuisance conditions arising from the hundreds of sanitary landfills located throughout the state. Although the Act was not intended to regulate sludges, such materials were specifically precluded from landfills. With the preclusion of petroleum sludges, paint sludges and a vast amount of other chemical and industrial wastes from land disposal sites, industry officials had to "scurry" around for a way to rid themselves of these materials.

In 1971 government investigators discovered approximately 3000 barrels of sludges in a landfill occupying a floodplain region in southeastern Michigan. The responsible industry was ordered to remove the drums from the floodplain site out of concern about possible contamination of the nearby flowing stream; even here, however, removal of the drummed sludges was predicated by concern for surface waters, not for the soils or underlying groundwaters.

Lack of control over proper handling of industrial sludge is evident in engineering plans prepared for wastewater treatment facilities. Such plans provide extensive detail of the component parts making up the treatment system. Drawings illustrate pump devices, chemical feed equipment, mixing apparatus, filtering devices, settling basins and dewatering components of the treatment system. But what do the same plans show for the solids extracted from the system? A close look at the end of the plan package where the vacuum filter equipment is drawn reveals a conveyor belt dropping the sludge into a truck. Finally, somewhere near the truck is a horizontal arrow with the words, "to disposal."

REGULATORY CONTROL

Most recent regulation of sludges can be attributed to new federal laws. In 1972 the Federal Water Pollution Control Act (PL 92-500) established levels of treatment and deadlines for meeting them, as well as penalties for violations. The Act established for the first time a national permit system requiring effluent limitations on concentration and mass limits. In the majority of cases, National Pollutant Discharge Elimination System (NPDES) permits mandated the upgrading of existing treatment plants or construction of new plants to provide higher levels of treatment and reliability.

PL 92-500, which aimed at providing a cleaner environment by removing greater masses of metals, nutrients and solids from the wastewater, did not take into account the disposal of sludges. The general aim of the Act was to clean up the water; it left the problem of sludge disposal control to be handled by state and local requirements. Regulatory control for sludge disposal is still the responsibility of state and local governments.

Later, with the adoption of the Clean Water Act of 1977 (PL 95-217), there was some added emphasis to regulate and control the disposal of sludges. Section 405 of the Clean Water Act requires the United States Environmental Protection Agency (EPA) to promulgate regulations governing the issuance of permits for the disposal of sewage. Further, in accordance with Section 402, EPA must develop and periodically publish regulations providing guidelines for the disposal and utilization of sludges. These regulations are to identify uses for sludges and to specify factors to be taken into account in determining the measures and practices applicable to each such use or disposal.

It is evident that further federal guidelines and regulations would place an even greater emphasis on sludge disposal. The NPDES permit provides the mechanism for such controls, and future permits are expected to place more restrictive limits on the quantity and kinds of toxic materials permitted to reach the general public.

As evident by the adoption of Act 64, P.A. of 1979 (the Hazardous Waste Management Act), Michigan feels it cannot wait for federally mandated regulations through the NPDES permit system. Legislation controlling and regulating most sludges of industrial origin is now in effect under the Toxic and Hazardous Waste Act, which took effect January 1, 1980.

VOLUME OF SLUDGE

At this time, there is no clearly established measure of the volume of industrial sludges generated in Michigan, but its vast automotive, chemical and paper industries are major generators of sludge materials. The petroleum, pharmaceutical, canning and brewing industries would contribute a sizable volume of sludges, but significantly less than do other sources.

A greater understanding of the magnitude of the sludge disposal problem began with the manifest system required by Act 64. For the first time, Michigan industries are now required to keep more specific records of their wastes, including sludges being generated by their manufacturing processes.

In 1979 the state of Michigan contracted with the Battelle Columbus Laboratories of Columbus, Ohio to identify the types and volumes of wastes generated by Michigan's industries. The main purpose of this study was to generate data on which to base the design for state-owned toxic and hazardous disposal facilities. Originally, it was conceived that such a facility would be owned and operated by the state. Now it appears unlikely that Michigan will proceed with the development of such a facility. State planners and policymakers believe that the state's role in disposal of sludges and toxic waste would best be limited to a regulatory responsibility of a facility developed by by private enterprise. Lack of interest on the part of private enterprise to develop adequate disposal facilities in the immediate future could bring about reevaluation of the state's role in this important area.

Where do the wastes go? It is not really known; however, if EPA studies are accurate and Michigan fits the norm, the following generally identifies various methods of disposal in this state:

1.　80% is disposed of on land or in nonsecure ponds, lagoons or landfills.
2.　10% is incinerated without proper controls.
3.　10% is managed in an acceptable manner.

EPA provides other rough estimates on the volumes of waste suspected of being generated in the states. These figures are as follows:

1.　10% of all wastes are hazardous.
2.　60% of the hazardous waste is considered to be in the form of liquids or sludges.
3.　60% of all hazardous waste is produced by the states of Texas, Ohio, Louisiana, Pennsylvania, Michigan, Indiana, Illinois, Tennessee, West Virginia and California.

Using the above percentages with an inventory conducted by the Michigan Department of Natural Resources in 1977, the volume of liquid, toxic and industrial sludges would be:

$$5024 \times 10^6 \text{ lb/yr of industrial residues}$$
$$\times \underline{0.10} \text{ (\% hazardous materials)}$$
$$500 \times 10^6 \text{ lbs per year}$$
$$\times \underline{0.60} \text{ (\% liquid or sludges)}$$
$$300 \times 10^6 \text{ lb/yr of sludges, or}$$
$$150,000 \text{ ton/yr}$$

Preliminary information being generated by the Battelle Laboratories study strongly suggests that our earlier estimates are in serious error. State officials now believe the quantity of industrial waste being generated by Michigan industries is three times greater than our original estimate. If this is true, more than 450,000 ton/yr of hazardous industrial liquids and sludges are being generated statewide.

This chapter presents a general discussion of sludges, their handling and disposal. It does not include a detailed technical evaluation of sludge characteristics or disposal methods, which are covered in subsequent chapters. Rather, this chapter will discuss the general types of sludges and the state's concern over ultimate disposal from a regulatory standpoint.

Sludges can be classified into three major categories:

1. Group 1 sludges: supernatant residues and/or leachate acceptable for groundwater and/or surface water discharge without extensive additional treatment.
2. Group 2 sludges: supernatant residues and/or leachate possessing pH values, iron concentrations, nutrient concentrations, etc., in excess of allowable levels for direct groundwater and surface water discharge without additional treatment and that do not contain toxic organic compounds or significant quantities of heavy metals.
3. Group 3 sludges: residues that contain significant quantities of toxic organic compounds or heavy metals that may become resuspended or soluble.

Group 1 sludges include fly ash slurries and certain air pollution scrubber streams where the sludges can be lagoon handled and the effluent normally can be monitored for total suspended solids and pH. This group also includes paper and brewery sludges, which are acceptable for direct land application.

Group 2 sludges represent such materials as lime sludges, iron precipitates and phosphate treatment sludges. The supernatant and/or leachate from systems handling the sludges may have to be pH adjusted as a minimum; they may even have to be redirected back to the facility for additional treatment prior to discharge to the waters of the state.

The third group of sludges is characterized by such residues as heavy metal precipitates, paint sludges, oily sludges, phenolic sludges and pharmaceutical and chemical manufacturing sludges. The discharges from systems handling these types of sludges will require extensive treatment of supernatants and/or leachates prior to even indirect discharge via a municipal sanitary sewer.

Unit operations employed prior to ultimate disposal of an industrial sludge are dictated by the individual characteristics of the sludge and the planned type of ultimate disposal for it. Sludges can be conditioned by inorganic chemical additives, such as ferric chloride or lime, or by the addition of polyelectrolytes. They can be thermally conditioned or conditioned by adding such materials as ash, dirt or paper pulp. They can be thickened, utilizing such operations as gravity thickeners, dissolved air flotation and centrifuges. There are many stabilization techniques, including anaerobic and aerobic digestion, chemical addition and chemical oxidation.

Sludges can be dewatered by material processes such as drying beds and lagoons or by mechanical means such as centrifuges, vacuum or pressure filters, and cyclones. They can be screened or gravity-filtered, heat-dried using flash injection and rotary kilns, or incinerated. Sludges can be chemically fixed to produce dry "innocuous" dirt-type material or encapsulated in a seamless polyethylene wrapper, if extremely hazardous. The listing of these historial unit processes for sludge handling is almost endless, and new processes are being marketed constantly. The various utilizations of these processes, singularly or in combination, would make an exhaustive list. As indicated earlier, choice of one of these processes is dictated by the characteristics of the untreated sludge and the planned ultimate disposal. Specific processes are usually decided on by the industry's in-house engineers or outside consultants, with little input by the regulatory agency.

On the other hand, the state regulatory agency has considerable input into the decisions involving the ultimate disposal of sludges. Both the state and federal governments support the concept of beneficial utilization of industrial sludges, which employs such approaches as sludge as a soil additive or conditioner, a direct or indirect energy source, and as a source of chemicals, raw materials, construction materials and animal feeds.

Certain sludges, generally those high in nutrients such as nitrogen and phosphate and low in metallics and undesirable organics, can be applied to crop lands and forest areas at a controlled rate to enhance the productivity of the soils. The rate of application is usually dictated by the levels of cadmium in the sludges, the types of crops and their ultimate use, and the biological concentration potential of the crop with respect to various contaminants. The physical limitation of the site, i.e., slope, soil porosity, etc., must also be taken into consideration and is generally evaluated through the use of a complete hydrogeological study defining groundwater aquifers and the direction of groundwater motion. Further, a complete determination of soil types and a complete soil analysis with and without sludge addition must be made. The sludge utilized for the experimental determination and the results of soil analysis must be handled by the same unit processes that will be employed at the generating industry, as many of the unit's sludge handling processes can concentrate undesirable trace contaminants in the untreated sludge to unacceptable levels in the treated sludges.

Unit treatment processes for sludge handling can be developed to utilize the Btu value of the sludges as a fuel in furnaces or incinerators. Methane or other volatile gasses generated during some sludge handling processes can be used as a fuel source rather than, or in addition to, the sludge itself.

Given the correct economic incentive, many sludges that contain metallics and/or organics could be used as a source of chemicals or raw materials. A case in point, involving the use of liquid waste rather than a sludge, is the use of waste-processing "pickle liquor" as a chemical source of iron in the removal of phosphate at municipal wastewater treatment plants. One Ann Arbor-area plater has been studying the reclamation of metals from an inactive sludge lagoon used by the company for many years. The major problem appears to be economic rather than technical. A Detroit-area steel company is utilizing waste sludges at a landfill located at one of its plants, with the ultimate intent of recycling the iron in the sludge.

Clearinghouses have been set up in some areas of the United States in an attempt to bring together sludge generators and potential reuse customers. The Southeastern Michigan Council of Governments (SEMCOG) has recommended that a similar cooperative group be established among the Detroit-area industries. Examples of sludge processed into animal feed are chicken manure being blended back into the chicken feed and the use of brewery waste solids as dairy cattle feed. The resultant materials from the chemical fixation of certain types of sludges using such processes as Chem-Pac$^®$ and Chem-Fix$^®$ have been used for containment dike construction, compacted dirt fill and highway construction fill.

The beneficial utilization of sludges should not be discounted because of past unsuccessful practices without first thoroughly investigating these potentials. The potential use of industrial sludges will be dictated by economics and by the technological advances needed to perfect the processes to separate desired materials within the sludges. In cases in which no beneficial use can be found, it can be disposed of at approved landfill sites. Landfilling involves only dry sludges, not liquids, and can be done in a dedicated industrial sludge disposal site or in combination with general refuse at an approved landfill. Site acceptability and proper handling depend on the chemical characteristics of the sludge.

Many sludges, such as those containing hazardous organics or high-level metallics, must be handled only in dedicated areas. These areas must be isolated from the groundwaters by natural formation or site engineering and equipped with collection and treatment systems for leachate control. Leachate from these areas must not be allowed to contaminate the potable groundwater supplies, as in Oakland County, or the surface waters, as they have in many areas of Michigan. A complete hydrogeological evaluation of the site must be made prior to any such sludge disposal to determine the potential for groundwater contamination and/or surface water pollution. The sludge must be analyzed for potential leachate contamination, utilizing the standard distilled water leaching test plus any modified testing that will predict the results of sludge idsposal at the selected site as accurately as possible.

An alternative to dry sludge landfilling is the land disposal of wet sludge. Several approaches to land disposal include spraying, flooding, ridge and furrow, spreading by tank truck and subsurface injection. Sites utilized for this type of disposal are designated strictly for sludges, not for codisposal with general refuse. These sites must be investigated thoroughly for both site topography and hydrogeological considerations.

PAPER INDUSTRY SLUDGES

All industrial sludges should not be judged as hazardous materials having a serious impact on the environment. Industrial sludges generated by Michigan's paper industries currently are being handled by several different methods. Some case histories to be discussed in later chapters will provide specific examples of sludge disposal procedures. The paper industry is a good example of diversity in sludge disposal methods. Several different approaches have been implemented with a minimum of difficulty and, in most cases, no adverse environmental problems.

Paper industry plants are scattered throughout the state of Michigan with no specific sector considered a predominant location. Disposal of sludges by the paper industry varies for each plant, with a host of factors considered significant in establishing a preferred method. As one might expect, there are many factors and considerations that determine the final method of disposal. In the case of the paper industry, a significant consideration must be given to economics. Since the paper industry is extremely competitive, economics must be weighed carefully in considering disposal of sludges.

The weather is a definite consideration in selecting a disposal method. A good number of paper mills utilize land application of sludges. This is an excellent method except during inclement weather conditions in the spring and fall when sludge cannot be placed on the land, prompting the need for good storage facilities. Some disadvantages arise with the use of storage facilities, such as the potential for odor nuisances and the need to handle the sludge more than once.

Although land disposal is a favored method, an Alpena Paper Mill is building a large dryer facility to remove moisture from the sludge to enable its utilization as a fuel source in plant operations. With rising fuel costs, it is conceivable that more consideration will be given to this method of disposal in future. Some disadvantages to this approach are environmental concerns, the high capital expenditure, operational expenses and fuel cost for the drying process. These disadvantages need to be balanced with the favorable features of low handling costs, low transportation costs and the lack of land disposal areas.

METAL INDUSTRY SLUDGES

Without question, sludges with the highest potential for creating environmental problems and those most difficult to handle are generated by the automotive and chemical industries. These sludges are classified as toxic, so require special handling.

Metal sludges used to be discarded in the most convenient location available to the industry. The various methods included disposal into lagoons on plant premises, in landfills, into public owned treatment works and by incineration. By today's standards, each of these practices is excluded from consideration because of its adverse environmental side effects.

In our opinion, hazardous sludges such as those from the electroplating industry should utilize land disposal only as a last resort. Regulatory agencies strongly prefer process changes so that the wastes are not created in the first place. More emphasis must also be given to reuse and recovery of waste so that these materials can be recycled as raw materials. After industry evaluates the full cost of disposal, it is believed a significantly higher level of reuse and/or recovery will be implemented.

CHEMICAL INDUSTRY

Sludges from the chemical industry represent a high level of concern because of their extreme toxicity and ability to cause or contribute to acute or chronic adverse effects on human health. Currently, these solids are being disposed of in many ways. There is some amount of recovery, some are being treated chemically for neutralization, some are solidified for land-fills, and some of the more toxic sludges are being incinerated.

While incineration is preferable for the very toxic organic chemicals, Michigan is very deficient in high-temperature incinerators that are equipped with pollution control systems and have adequate retention times. There are few incinerators available in the state to handle contaminated sludges. As a rule, companies with facilities now in operation will not accept contaminated sludges for commercial disposal.

Where companies have attempted to use existing incinerator facilities to rid society of some contaminated sludges, citizens objections have overridden such action. A case in point is a cement plant in the Detroit area that sought a permit to burn polychlorinated biphenyls (PCB) in its kiln as part of a routine cement production operation. After public hearings and a two-year effort to obtain a permit, the company finally abandoned any hope of obtaining it. This is most unfortunate because state officials believed that no better method could be obtained to handle the thousands of pounds of contaminated materials now in storage.

One chemical firm in the state had an extraordinary perception of the eventual difficulty in the disposal of toxic and hazardous materials and built a high-temperature rotary kiln. This Midland-based company is one of a very few companies that has the ability to rid itself of hazardous sludges from its manufacturing processes. Its incinerator has been in operation for approximately seven years.

PRETREATMENT

Implementation of the state's pretreatment requirements is expected to have a varied and significant impact on Michigan's industries. It is not known at this time whether the program will create more or less sludge or exactly what will be the eventual disposal method for the sludge produced. On the surface, it would appear that pretreatment will generate more sludge. This supposition may not be true of the municipal sludge program. A review of existing records reveals that more than 200 municipalities have sludges in the hazardous waste category because of excess chemicals and metals. If pretreatment eliminates hazardous substances from the POTW, then conceivably large volumes of previously contaminated municipal sludge would be eliminated from the hazardous waste category. If one assumes that the industrial sludge was a small fraction of the total volume of sludges produced by the municipality, then the overall volume could be reduced significantly.

The volume of sludge generated depends on industry's approach to the problem. If new processes involve recycle or recovery systems, then the production of sludge can be eliminated or reduced significantly. It is too soon to determine what effect, if any, recycling or recovery will play in terms of volume generation. Industry will need to evaluate capital expenditure and the operational costs for various alternatives before choosing a preferred method of disposal.

Implementation of the program to regulate electroplating waste in municipal systems will see Michigan industries involved in a flurry of activity to comply with new regulations. EPA is providing new facilities with a three-year time limit to eliminate heavy metals in municipal systems. These regulations will create an undetermined, but significant, volume of new industrial sludges for disposal. Although ultimate disposal of these materials presently is unknown, it is anticipated that considerable quantity will go to municipal landfill sites. Sludge not disposed of in this manner will be incinerated, recycled, or recovered. But most will probably find its way to a landfill and will require the generator to adhere to the regulations of Act 64, P.A. of 1979. The following are some of the important provisions of the Act as they apply to industries producing hazardous sludges. Reviewing the Act raises the following questions regarding disposal of sludges from manufacturing processes:

1. When Does the Act Go into Effect?
The Act went into effect January 1, 1980. The rules by which the Act will be administered have not been finalized; however, it is anticipated that all generators will have initiated the manifest system required by the Act.

2. What Wastes are Considered Hazardous?

Flammable wastes, explosive wastes, waste gases under pressure, toxic wastes, metal wastes or infectious wastes are considered hazardous wastes under the Act. Standard industrial codes are stated in the Act to identify hazardous wastes sources.

3. Who Must Comply with the Act?

The Act states that "generator" means any person, federal government, facility or installation, education or health care facility, or any public or private entity, whether for profit or nonprofit, whose activity results in generation of waste that is or may become hazardous.

4. Who Determines Whether the Waste is Hazardous?

The Act provides a list of wastes by the standard industrial classification number of the most common types of generators. The generator may submit a demonstration for his particular waste if he feels it does not possess hazardous characteristics.

5. Who has Full Responsibility for Proper Disposal of Hazardous Waste?

The generator is responsible for proper disposal of his waste and is not absolved of this responsibility until he receives a certificate of disposal.

6. What Should a Generator Do to Maintain Compliance?

He should immediately initiate the manifest system. Next, he should obtain a hauler legally licensed in accordance with Act 36 to transport the waste to a licensed disposal site. After acceptance of the waste, the generator is given a certificate of disposal that will absolve him of further responsibility for the waste.

7. Does the Generator Need a License Under the Act?

In accordance with the Act, the generator does not require a license to haul his own company's waste to a disposal site. He must have only a vehicle license and be able to show proof of financial responsibility.

8. What Does the Manifest Indicate?

It identifies the licensed hazardous waste hauler and the permitted disposal facility to which waste is delivered. It assures that proper containerization, labeling and placards are used for the waste. The manifest also provides the disposal facility with information necessary for proper disposal.

9. How Long Must a Generator Retain a Manifest?

The generator shall retain copies of the manifest records for a review by the Department for a three-year period. The Department may request the submission of records at any time and such information must be submitted within 30 days after receipt of request.

10. What Limitation will be Placed on Land Disposal of Sludge?

Many sludges will be acceptable for disposal on the land. Once the pretreatment program is implemented and in effect, it is anticipated that most municipal sludges will be acceptable for land disposal. Wastes containing metals less than ten times the United States Public Health Service Drinking Water Standards (1962) will be acceptable for land disposal.

11. How Will Act 64 be Enforced?

No firm enforcement procedures have been developed at this time. However, generators can expect an occasional audit of handling procedures. Most reviews will take place when the state investigators conduct comprehensive evaluation inspections and sampling surveys to establish compliance with the NPDES Permit.

12. What are the Penalities?

Violation of Act 64 has a potential for pollution of the groundwaters and surface waters of the state. A violation of the state's water quality standards under Act 245 of PA 1929, as amended, is subject to $25,000 for each violation. There are provisions in the law to issue a $50,000 fine for second offenses.

13. Can a Generator Declare a Sludge Hazardous?

Yes. The generator may have a small volume of sludge containing various unknown characteristics and may wish to declare it hazardous. This would be done to avoid the costly expense of evaluation. Generators may declare their waste hazardous and dispose of them in accordance with the Act; however, the generator must provide sufficient information about the waste to assure proper management.

14. Will the List of Hazardous Wastes be Revised?

Yes. Whenever information becomes available indicating that a waste is hazardous, it will be evaluated and placed on the list. On the other hand, materials may be removed from the list. If there is enough substantiating data developed to indicate a substance is not toxic, then rule change procedures can be initiated to delist the waste.

15.   What Type of Sludges are Classified as Hazardous?

Paint and latex sludges, electroplating sludges, lubricating sludges, acid sludges, wool-fabricating dye sludges, knit-fabric dying sludges, wool-scouring sludges, chrome-bearing sludges, numerous types of filter cakes, tars and still-bottom petroleum refining sludges, API separator sludges and all of the numerous heavy metal-type sludges.

16.   What Other Regulations Does Act 64 Provide?

The Act provides permits for disposal sites and a separate permit for construction and operation of facilities. Further, the Act establishes monitoring requirements, operation and closure procedures, spill prevention plans and requirements for incinerator performance.

17.   Will a Landfill Accept any Hazardous Waste?

No. Flammable, reactive and volatile wastes will not be placed in a landfill. Sludges must be of a nonflowing consistency before they can be accepted for disposal in a landfill.

18.   Will the Generator be Required to Have a Short-Term Storage Permit?

No. However, a person shall maintain good inventory records and make periodic visual inspection of facilities. He shall take appropriate steps to prevent the loss of waste to the groundwaters or surface waters.

19.   Is There a Limit on Short-Term Storage?

Yes. The quantity of waste stored at any location shall not exceed 100,000 pounds, unless otherwise authorized in writing by the Department.

20.   What are the Reporting Requirements for a Short-Term Storage?

The person who has a short-term storage shall submit to the Department the type and maximum quantity of hazardous waste stored and volume within 30 days of implementation of the storage.

SUMMARY

For years, environmental control specialists throughout the United States paid little attention to sludge disposal because they were unaware of the potential problems it could create. Just recently have there been examples of the tragic consequences of this lack of knowledge with the seepage of chemicals from the ground at Love Canal in New York, causing two hundred families to leave their homes. Examples of similar problems closer to home include the burial of hundreds of drums containing sludges at two locations in northern Oakland County. Hooker Chemical's burial of waste in Muskegon County and Velsicol Chemical Company's burial of waste in the Alma-St. Louis area are further examples of poor judgment by industry in the disposal of chemicals and sludges.

Past problems with disposal of sludges and chemicals at landfill sites have made it virtually impossible to locate new disposal facilities in the state. Establishing a disposal facility site in the lower, more populous region of the state where such facilities are in greatest demand is especially difficult. The public does not want such a facility anywhere nearby. More than just the facility, the public opposes the noise problem associated with the operation and expresses much concern over the nuisances arising from dust and odors. Location of such a facility near a property owner also raises concern over land values. The end result is that nobody wants the site next door.

Obviously, if sludges and hazardous wastes are to be managed successfully in the state, we will need sites to store and dispose of these materials. The goal of the Michigan Department of Natural Resources is to earn the confidence of the public by establishing high standards and to take quick and decisive enforcement action against violators. Once it has been proven to the general public that adequate safeguards are established and carried out, the Department will gain the trust of Michigan citizens and have less difficulty in locating needed disposal sites.

CHAPTER 2

MUNICIPAL WASTEWATER
TREATMENT PLANT SLUDGE

Richard T. Sprague, Specialist
Onland Waste Treatment
Michigan Department of Natural Resources
Lansing, Michigan

Municipal wastewater sludge disposal is an area of growing concern. Michigan wastewater treatment plants produce more than 1000 dry ton/ day of sludge, a quantity that is expected to increase by about 50% by 1985 [1]. Further, septic tank pumpings produce approximately 100 dry ton/day of material that is very similar to residential sewage sludge [2]. Finally, industry produces approximately 2000 dry ton/day of sludge from biologically derived processes [3]; again, many of these materials are similar to domestic sewage sludges. As the quantity of sludge increases, so does the number of regulations attempting to control this material, as well as other potentially toxic or hazardous wastes. This has created an atmosphere of confusion, concern and misunderstanding within the sludge disposal industry.

This chapter discusses present and future regulations on sludge disposal in the state of Michigan. As other authors in this book discuss incineration, energy recovery and landfilling, this discussion will be limited to the regulation of land application, and land disposal or reclamation projects.

LAWS, RULES AND REGULATIONS

Three state actions are having an effect on sludge disposal within the state of Michigan. The first—the proposed Groundwater Rules—have been proposed by the Michigan Department of Natural Resources (DNR) to protect and regulate discharges into aquifers of the state. The second—the Solid Waste Management Act (SWMA) (1979, P.A. 641)—regulates disposal of all solid wastes. The third—the Hazardous Waste Management Act (HWMA) (1979, P.A. 64)—controls transportation, disposal and recovery of hazardous wastes. These three sets of regulations have stimulated discussion at the state level similar to that encountered at the federal level [4].

Groundwater Rules

In 1976 the Michigan Department of Natural Resources charged a committee with the development of rules intended to protect public health and welfare by maintaining the quality of the groundwaters of the state. This committee represented the interests of public health, agriculture and transportation, as well as natural resources. The rules, proposed under the Michigan Waters Resources Commission Act (1929, P.A. 245), do not allow for degradation of groundwater quality in usable aquifers. The term "usable aquifer" was defined in these rules as an aquifer or a portion thereof that is of a quality and quantity satisfactory for use as an individual, public, industrial or agricultural water supply. Further, the rules define the hydrogeological study required prior to permitting a discharge into the groundwaters, as well as establish groundwater monitoring requirements for new and existing groundwater discharges. Finally, the proposed rules establish a variance procedure.

These rules do not allow discharge of any material onto or into the ground if the discharge could result in degradation of the usable aquifer. However, the rules specifically exempt some types of discharge to the surface of the soil. These exemptions include such discharges as controlled application of dust-suppressant or deicing chemicals. They also specifically exempt controlled application of agricultural or silvacultural chemicals. It was the intent of the rules committee to include within this exemption sludge or wastewater applications that result in controlled application of agricultural nutrients contained in the waste material. In Chapter 9 Dr. Jacobs will discuss controlled application of agricultural chemicals. For specifically exempted discharges, groundwater monitoring is not required; however, groundwater monitoring may be desirable as further demonstration of protection of usable aquifers. While these rules were approved by the Michigan Water Resources Commission in 1977, they still await approval by the state Legislature.

Solid Waste Management Act

The SWMA was intended to protect the public health and the environment by ensuring proper disposal, and by encouraging recycling, of solid wastes. This Act defines sludge from municipal and industrial waste treatment plants as a solid waste. In doing so, the Act is consistent with the Federal Resource Conservation and Recovery Act (RCRA). The Act provides specific requirements for solid waste disposal facilities, including site bonding, submission of engineering plans and specification, issuance of construction permits and operating licenses, and recordkeeping.

Some of these provisions are not relevant to land application of sewage sludge; in fact. several of these provisions would discourage utilization of sludge on agricultural land. This is in direct conflict with a stated aim of the Act—recycling. Consequently, policies exempting certain municipal sludge disposal practices and facilities from provision of the SWMA were developed by Michigan's DNR.

Figure 1 illustrates this intent. Where possible, regulation of sludge under several acts has been avoided; however, the Decision Tree also clearly defines disposal methods for which regulation under several acts is required. This schematic serves as a basis for regulations under development.

Hazardous Waste Management Act

In 1978 the Michigan's DNR proposed legislation aimed at controlling hazardous wastes. This proposal would have created a state-owned facility designed to recover, chemically or thermally destroy, or biologically treat hazardous wastes. Landfilling would have been reserved for a limited number of hazardous wastes; however, landfilling would have also been designed into the facility. In rejecting this proposed legislation, the legislature established a working group to develop alternative legislation. This working group was not a typical legislative committee in that more than 50 legislators had input into the development of the HWMA. The Act was passed by acclamation in 1979.

The intent of this Act was to plug all regulatory loopholes to ensure that all hazardous wastes produced within the state were treated or disposed of adequately. The Act does not include or exclude any specific disposal options. It defines hazardous waste as any waste which, *if mismanaged*, will cause or may cause a substantial potential or present hazard to human health or the environment.

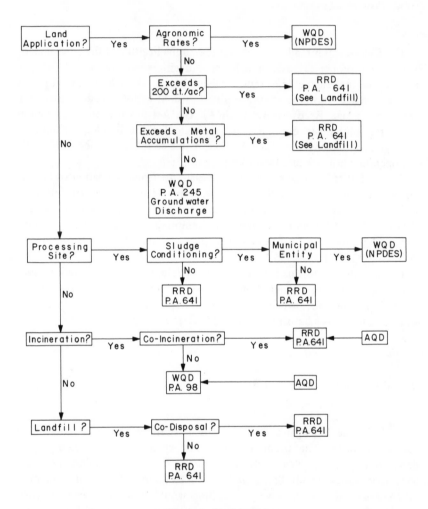

Figure 1. Decision tree.

Provisions of the Act

The HWMA placed specific controls on the transportation, storage and disposal of all wastes that are hazardous because of a flammable, corrosive, reactive, gaseous, toxic, metallic or infectious constituent. The Act provided a dual review of new hazardous waste disposal facilities, which incorporates both technical and sociopolitical evaluation. Following receipt of a complete application for construction permit, the director of the Michigan Department of Natural Resources must reject or recommend approval of the permit application within 120 days. Final approval of a permit application is granted or denied by a site approval board. The site approval board is created if the director has not rejected a permit application within 75 days, or if the director has recommended approval within this period. This site approval board must convene within 130 days of receipt by the director of a complete permit application, and has a maximum of 120 days to approve or reject an application. The site approval board consists of three representatives of state government; two representatives of the county; and two representatives of the city, town or township in which the facility is proposed; and a chemical engineer and a geologist, both of whom are faculty members at a state college or university. Therefore, approval or rejection of a permit application may require eight months.

Following receipt of a complete application for operating license, the director must reject or approve the permit application within 90 days. Under the Act, no local ordinance can supersede an approved operating license. Among other requirements, operating license approval requires a $5,000,000 bond for accidental spill, and a $5,000,000 bond for slow release of hazardous waste or hazardous waste constituents from the facility.

Municipal sludge is not exempted from the provisions of the Hazardous Waste Management Act if it is determined to be hazardous. This is consistent with the Act and is a valid environmental protection position because mismanagement of sludge could result in environmental damage. Most municipal sludge will probably meet the criteria of hazardous waste as defined in the current draft of the rules. These criteria were developed by technical committees to define the constituents and the concentrations of these constituents in a waste that pose a real danger to public health or the environment if mismanaged. They were developed neither with the intent to include, nor with the intent to exclude, sludge.

In late 1979 six sludges produced at wastewater treatment plants within the state were sampled. These treatment plants were chosen on the basis of varied industrial inputs and included two communities with essentially no industrial input. The levels of organic contaminants detected in the sludges would classify each as hazardous wastes. Apparently, industrial pretreatment programs will not reduce all of the organic and inorganic constituents to levels that would exempt sludge from the provisions of the HWMA.

The Problem

Municipal sludge is a high-volume waste. Disposal of municipal sludge in hazardous waste disposal facilities will strain the capacity of these facilities. Moreover, sludge is produced throughout the state, while hazardous waste disposal facilities are likely to be few and isolated. The site approval board and financial capability requirements imposed by the HWMA will limit the number of facilities throughout the state and impose a time delay on their development. Landfilling of sludge for 20-year planning period will require a total of 68,000 ac-ft of disposal site. This is equivalent to 5.3 mi^2 at a depth of 20 ft, assuming 20% solid sludge filter cake bulked to 50% solids prior to landfilling. The liquid present in most sludges exacerbates leachate production in landfills. Finally, low levels of organic or inorganic constituents in sludge may contribute to leachable concentrations of these constituents in a landfill. Disposal of sludge in licensed hazardous waste landfills will result in an increase of three to ten times present costs for municipal sludge disposal. The exact cost depends on the number of landfills because it primarily reflects transportation distance; however, present landfilling sites and techniques are inadequate in many communities; therefore, some increase above present cost is expected in any case.

Management of sludge as a hazardous waste essentially will eliminate land application as a management option. This is in direct conflict with our current policy, which encourages recycle. Site approval board and financial capability requirements of the Act will impose unworkable constraints on the use of privately owned farmland. Site approval board review would require commitment of fields and projection of crops by private landowners years in advance. These private landowners are unlikely to accept designation of their farmland as a hazardous waste disposal site. Land application of all sludge produced in the state would require approximately 140,000 acres, or 2.1% of the land under cultivation in 1978. Not all Michigan sludges are compatible with land application, at least on agricultural land. Land application of sludges presently compatible with agrucultural use would require approximately 50,000 acres. Disposal of sludge on agricultural land is also going to be more expensive in the future for most municipalities. We estimate increases in cost of two to four times present cost (based on limited data) for proper handling of these sludges. Land application may be the preferred treatment for low levels of many organic compounds. That is, microbial degradation and incorporation into soil organic matter will render immobile and inert such compounds as phenols [5] and phthalates [6], at least in low concentrations.

Forest application suffers many of the same problems as agricultural application; however, it reduces the potential for movement of toxicants into the human food chain. Utilization of sludge on forest land is limited due to lower nutrient assimilative capacity and to more difficult application in forested areas.

Incineration of sludge may be possible only in large municipalities. Economics of sludge incineration are likely to change in future because of both the provisions of the Hazardous Waste Management Act and increasing energy costs.

The Options

The problems created by these definitions are similar to those faced by the U. S. Environmental Protection Agency (EPA) in development of hazardous wastes and sludge disposal regulations. The options available include the following:

1. All sludge disposal facilities and practices would be licensed under the Hazardous Waste Management Act.
2. Selected sludge disposal facilities or practices regulated under other rules would be exempted. Disposal facilities or practices not exempted would be regulated under the Hazardous Waste Management Act.
3. All sludge disposal facilities or practices would be exempted by the Legislature.

LAND APPLICATION AND LAND DISPOSAL

In response to this dilemma, the DNR is developing regulations that will control disposal or utilization of municipal sludge on land. These regulations will also control land disposal or utilization of similar waste products, such as septage and industrial sludges. Specific regulations will control the following;

1. The first will be the agricultural utilization of *recyclable* sludges, which will balance sludge nutrient value against the risk attributable to potential toxicants contained in the sludge. When these rules and criteria defining recyclable are promulgated, this practice will be exempted from HWMA control.
2. Land disposal or reclamation utilizing recyclable wastes is the second.
3. Third is land treatment of hazardous, nonrecyclable wastes.

The regulations controlling septage disposal will allow delegation of authority to certified local health departments. However, the rules will establish the criteria by which a health department will evaluate an application for disposal permit.

Agricultural Utilization

In 1979 the DNR developed a strategy for the regulation of municipal sludge used on the agricultural land that forms the basis of rule development. This package included guidance on isolation distances (Table I), acceptable crop types (Table II) and acceptable levels of potential toxicants (Table III);

Table I. Geology, Isolation and Stabilization
Required for Surface or Subsurface Application

Factor	Application	
	Surface	Subsurface
Slope	0-6%	0-12%
Depth to High Water Table	>3 ft	>3 ft
Isolation		
Wells	200 ft	100 ft
Residences	500 ft	100 ft
Surface waters	200 ft	50 ft
Roads	200 ft	25 ft
Stabilization		
Unstabilized	U^a	M
Aerobic digestion	A^b	A
Anaerobic digestion	M^c	A
Liming to pH 12	A	A
Dry heating	M	A
Wet air oxidation	M	A
Composting	A	A

^{a} U = unacceptable.
^{b} A = acceptable.
^{c} M = marginally acceptable.

however, the key elements contained in this strategy included prior approval of a sludge disposal plan—Program for Effective Residuals Management (PERM)—and the keeping of records on disposal.

Program for Effective Residuals Management

Facilities that have an existing groundwater or surface water discharge have, or will have, a permit requirement for a PERM. This report will present the following, in detail, for the upcoming sludge disposal year:

1. production and treatment of sludge;
2. storage capacity to be utilized when disposal is not feasible;
3. disposal sites that will be utilized, including hydrogeological survey or agricultural management plan;
4. monthly budget of the sludge produced, stored and disposed;
5. monitoring of the environment and sludge; and
6. contingency plans, including backup disposal sites and cleanup procedures following a spill.

This report must be submitted to the Department of Natural Resources for approval and must be updated annually. Any unapproved disposal will be a violation of the existing discharge permit.

Table II. Stabilization Techniques Required for Present or Future Crops

| | | | Stabilization | |
| | | | | |
Crop	None	Digestion	1. Liming 2. Dry Heating	1. Composting 2. Wet Oxidation
Present				
Unprocessed fruit and vegetables	U[a]	U	U	U
Processed fruit and vegetables	U	M[b]	A[c]	A
Grain	M	A	A	A
Hay, haylage	U	A	A	A
Pasture	U	M	A	A
Nonfood	A	A	A	A
Future				
Unprocessed fruit and vegetables	3 yr	3 yr	1-3 yr	1 yr
Processed fruit and vegetables	3 yr	1 yr	A	A
Grain	1 yr[d]	A	A	A
Hay, haylage	1 yr	A	A	A
Pasture	1 yr	1 yr	A	A
Nonfood	A	A	A	A

[a] U = unacceptable
[b] M = marginally acceptable
[c] A = acceptable for present crop; no isolation period required.
[d] Less when crop will be used only for direct animal consumption.

Sludge Disposal Recording

Figure 2 presents the monthly operating report developed to conform with highly managed agricultural utilization of sludge. This form will provide a permanent record on the quantity of sludge delivered to, and utilized on, a particular site each month. It will also present the fertilizer requirements of crops, and site characteristics, such as soil pH and cation exchange capacity (CEC). It will provide typical analytical data on the sludge produced during the disposal period, or the average quality of the sludge. Finally, the form will present the total accumulation of both nutrients and potential toxicants on a monthly, annual and historical basis. This monthly operating report will allow ready evaluation of existing sludge disposal sites and will provide a permanent record that can be utilized for environmental and crop quality monitoring.

Table III. Sludge Quality

		Use		
Parameter	Nonrestricted	Slightly Limited	Severely Limited	Restricted
Cadmium	5	5-25	25-125	125
Chromium	50	50-1000	1000-5000	5000
Copper	250	250-1000	1000-2000	2000
Lead	250	250-500	500-2000	2000
Mercury	2	2-5	5-10	10
Nickel	25	25-200	200-1000	1000
Zinc	750	750-2500	2500-5000	5000
Selenium	10	10-40	40-80	80
Molybdenium	10	10-20	20-50	50
Arsenic	100	100-500	500-2000	2000
Polychlorinated Biphenyls (PCB)	1	1-10	10-50	50
Polybrominated Biphenyls (PBB)	1	1-10	10-50	50
Other Organics	NS[a]	NS	NS	NS

[a] NS = not sufficient data.

Land Disposal or Land Reclamation

Applications of a recyclable sludge that exceeds the nutrient requirements of agronomic crops will be defined as land disposal or land reclamation. Land disposal and land reclamation have been demonstrated elsewhere to be viable sludge disposal options [7]. Where the hydrogeological conditions are appropriate, the regulations will provide for issuance of a groundwater discharge permit. This permit will require approval of a PERM prior to disposal or reclamation, and maintenance of records, both as outlined above.

Hazardous Waste

A separate set of regulations has already been drafted to control land treatment of hazardous wastes. These rules will be promulgated under the Hazardous Waste Management Act.

State of Michigan
Department of Natural Resources

SLUDGE DISPOSAL SHEET

_____ (Month) _____, 19____

_____ (Municipality) _____, Michigan

(Superintendent's Signature) _____

_____ (site) _____ ac.

Acres This month _____

CROP AND SOIL DATA

Present Crop _____

Projected Crop _____

CEC _____ meq/100g

pH _____ S.U.

Fertilizer Recommendations

N _____ lb/ac

P_2O_5 _____ lb/ac

K_2O _____ lb/ac

Acceptable Metal Accumulations

Pb _____ lb/ac

Zn _____ lb/ac

Cu _____ lb/ac

Ni _____ lb/ac

Cd _____ lb/ac

Cd 2 lb/ac/yr

REMARKS:

DATE	SLUDGE APPLIED				Nitrogen			Total Phosphorus	Potassium	SLUDGE ANALYSIS AND SOIL LOADING RATES													
	Gallons or Cubic Yards	% Solids	% VS	Dry Tons per ac	TKN %	NH₄ %	NO₃ %	AVAN lb/ac	TP %	K %	Lead Pb		Zinc Zn		Copper Cu		Nickel Ni		Cadmium Cd					
											mg/kg	lb/ac	mg/kg	lb/ac	mg/kg	lb/ac	mg/kg	lb/ac	mg/kg	lb/ac			

Average

T This Month
O
T This Year
A
L Cumu-
S lative

SLUDGE DISPOSAL SHEET INSTRUCTIONS

Each sludge disposal site should be reported separately. A change of crop or soils, as well as a change of landfill, constitutes a change of sludge management, and should be reported separately.

Sludge Applied—Report the volume of sludge delivered to a site on each day in gallons or cubic yards. Also, report percent total solids and percent volatile solids on the days analyzed. Calculate the monthly application rate, in dry tons/mo, according to the appropriate formula:

Dry Tons/mo. = _____ gal/mo x _____ % solids x 0.0000425

or   Dry Tons/mo. = _____ cu. yds/mo. x _____ % solids x 0.008425

Sludge Analysis—Report the analysis of sludge in percent for N, P and K, and in mg/kg for metals. (Note: 10,000 mg/kg = 1.0%, 1000 mg/kg = 0.1%.) Report the average analysis for this month, or for the most recent samplings, near the bottom of the sheet. The frequency of sludge sampling was determined in your Sludge Management Plan.

Soil Loading Rates

Nitrogen

Ni = NH_4^+-N ___ ___.___% + NO_3^--N ___.___% = ___ ___.___%

No = TKN ___ ___.___% − NH_4^+-N ___.___ ___% = ___ ___.___%

Available Nitrogen (AVAN)

Ni = ___ ___.___% x 20 = ___ ___.___ lb/ton

No = ___ ___.___% x 4 = ___ ___.___ lb/ton

+ _____

AVAN = ___ ___.___ lb/ton

AVAN = ___ ___.___ lb/ton x ___ ___.___ton/ac =

___ ___.___ lb/ac

Report this value at the bottom of the sheet for monthly totals of AVAN.

Phosphorus and Potassium

___ ___.___% (TP or K) x ___ ___.___ tons/ac x 20 = ___ ___ ___ lb/ac

Metals

_____ mg/kg metal x ___ ___.___ tons/ac x 0.002 = _____ lb/ac

Crop and Soil Data.—Transfer this information from soil test results. Include present ground cover as well as the projected crop, cation exchange capacity (CEC) and nutrient recommendations. Calculate and record the metal loading limits based on the following factors and CEC:

Metal	Factor	
Pb	100	
Zn	50	*25 where pH control of 6.5 is assured.
Cu	25	
Ni	10*	
Cd	1	

CEC ___ ___.___ meq/100g x ___ ___ (factor) = ___ ___ ___ lb/ac

CONCLUSIONS

Regulations currently are being drafted to control land application of municipal and industrial sludge, and septage. These regulations were to receive their first public airing before August 1980 and are based on widely distributed departmental guidance and strategy.

REFERENCES

1. "Municipal Wastewater Sludge Management," staff report to Michigan Environmental Review Board (1979).
2. "Alternatives for Small Wastewater Treatment Systems," Vol. 1, U. S. Environmental Protection Agency, EPA-625/4-77-011 (1977).
3. "Report on a Survey of Industrial Waste Management to Michigan Department of Natural Resources," Battelle Columbus Division, Battelle Laboratories, Columbus, OH (1980).
4. Weddle, B. R. "Impact of EPA Rulemaking on Sludge Utilization and Disposal," in *Eighth Nat. Conf. on Municipal Sludge Management,* (Silver Spring, MD: Information Transfer, Inc., 1979), pp. 7-9.
5. Overcash, M. R., and D. Pal. *Design of Land Treatment Systems for Industrial Wastes* (Ann Arbor, MI: Ann Arbor Science Publishers, Inc., 1979), pp. 233-244.
6. Inman, J. C. "Sewage Sludge Applications to Soils: Laboratory and Field Studies of Heavy Metal Extractability and Carbon and Nitrogen Transformations," Ph.D. Thesis, Purdue University, West Lafayette, IN (1979).
7. Sopper, W. E., and S. N. Kerr, Eds. *Utilization of Municipal Sewage Effluent and Sludge on Forest and Disturbed Land* (University Park, PA: The Pennsylvania State University Press, 1979).

CHAPTER 3

MANAGEMENT OF WATER TREATMENT PLANT SLUDGES

David A. Cornwell
> Assistant Professor
> College of Engineering
> Michigan State University
> East Lansing, Michigan

Garret P. Westerhoff
> Vice President
> Malcolm Pirnie, Inc.
> White Plains, New York

Management of water treatment plant sludges has become an increasingly difficult and expensive requirement. The most abundant types of water plant sludges are produced at chemical coagulation and lime softening plants, both of which account for about 95% of all water treated. Coagulant plants represent about 70% of all treatment plants; consequently, the management of coagulant sludges has been emphasized in this chapter.

This chapter has been divided into five general categories: (1) production and characteristics of sludge; (2) minimization of sludge generation; (3) chemical recovery; (4) treatment alternatives; and (5) ultimate disposal.

SLUDGE PRODUCTION AND CHARACTERISTICS

In water treatment plants, sludge is most commonly produced in the following treatment processes: presedimentation, sedimentation and filtration (filter backwash).

31

Presedimentation

When surface waters are withdrawn from water courses containing a large quantity of suspended materials, presedimentation prior to coagulation may be practiced to reduce the accumulation of solids in subsequent units. The settled material generally consists of fine sands, silts, clays and organic decomposition products.

Lime Sedimentation Basin

The residues from softening by precipitation with lime $(Ca(OH)_2)$ and soda ash (Na_2CO_3) will vary from a nearly pure chemical to a highly variable mixture. By simplifying the softening reactions to one step, the production of the pure chemical sludge is illustrated:

$$Ca(HCO_3)_2 + Mg(HCO_3)_2 + 3Ca(OH)_2 = 4CaCo_3(s) + Mg(OH)_2(s) + 4H_2O$$

Theoretically, each mg/1 of calcium hardness removed produces 2 mg/1 of $CaCO_3$ sludge, and each mg/1 of magnesium hardness removed produces 2.6 mg/1 of sludge. The theoretical sludge production can be calculated as

$$S \quad = \quad (2Ca + 2.6\ Mg)\ Q\ (8.34)$$

where S = sludge produced lb/day,
           Ca = mg/1 as $CaCO_3$ of calcium hardness removed,
           Mg = mg/1 as $CaCO_3$ of magnesium hardness removed, and
           Q = plant flow, mgd.

A survey by the state of Ohio found actual sludge production at lime softening plants to be 1.75 times the theoretical values [1]. However, many of the plants surveyed were softening surface supplies, which produce a highly variable material. In this case, softening is often carried out in conjunction with chemical coagulation and, therefore, may contain large quantities of silts, clays and precipitated metal coagulants.

Coagulant Sedimentation Basin

Aluminum or iron salts are generally used to accomplish coagulation. The chemistry of the two salts is similar, so alum alone is referred to below.

The reactions of alum with water are complex, leading to the formation of insoluble aluminum hydroxide species. The reaction is often written as follows:

$$Al_2(SO_4)_3\ 14H_2O + 6HCO^{-3} = 2Al(OH)_3(s) + 6CO_2 + 3SO_4^{-2} + 14H_2O$$

The simplified reaction shows that as alum is added to the water it will react with the alkalinity and produce aluminum hydroxide. Figure 1 shows a theoretical solubility diagram for the precipitation of aluminum hydroxide, which is the predominant species that would exist if equilibrium were reached in the coagulation process. However, in the actual treatment process, equilibrium is not reached and various intermediates are formed. Figure 2 shows some of the reactions involved in the hydrolysis of Al^{3+}. Aluminum ion (reaction 1a on Figure 2) does not exist alone in water, but rather is bound to six molecules of H_2O. Since it is not known which of many possible species forms, it is difficult to estimate the quantity of chemical sludge produced as a result of aluminum addition. However, in the pH range of 6-8, in which most plants coagulate, probably the insoluble aluminum hydroxide complex of $Al(H_2O)_3(OH)_3$ predominates. This species results in the production of 0.44 pounds of chemical sludge for each pound of alum added.

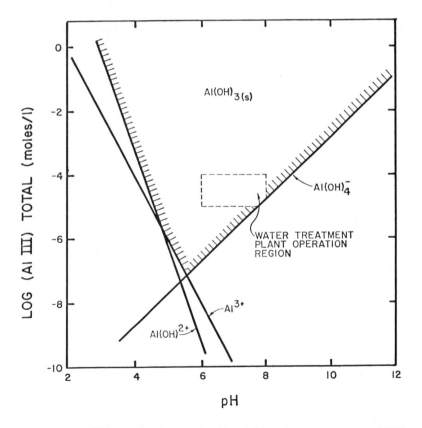

Figure 1. Aluminum hydroxide solubility diagram.

Ligand Exchange Reactions

1. a. $[Al(H_2O)_6]^{+++} + H_2O \rightleftharpoons [Al(H_2O)_5 \, OH]^{++} + H_3^{\,+}$

   b. $[Al(H_2O)_5OH]^{++} + H_2O \rightleftharpoons [Al(H_2O)_4 \, (OH)_2]^{\,+} + H_3O^{\,+}$

   c. $[Al(H_2O)_4(OH)_2]^{\,+} + H_2O \rightleftharpoons [Al(H_2O)_3 \, (OH)_3] + H_3O^{\,+}$

2. $\quad [Al(H_2O)_3(OH)_3] + X^{=} \rightleftharpoons [Al(H_2O)_4 \, X]^{\,+} + 2H_2O$

3. $\quad [Al(H_2O)_4(OH)_2]^{\,+} + X^{=} \rightleftharpoons [Al(H_2O)_4 \, X]^{\,+} \dagger \, 2OH^{-}$

(X represents a divalent ion)

Olation Reactions

4. a. $2[Al(H_2O)_5OH]^{++} \rightleftharpoons$ $(H_2O)_4Al \overset{OH}{\underset{OH}{\diagup\diagdown}} Al(H_2O)_4^{\,+4} + 2H_2O$

   b. $[(H_2O)_4Al(OH)_2Al(H_2O)_4]^{+4} \rightleftharpoons [(H_2O)_4Al(OH)_2Al(H_2O)_3OH]^{+3} + H_3O^{\,+}$

   c. $[(H_2O)_4Al(OH)_2Al(H_2O)_3OH]^{+3}$
      $+ [Al(H_2O)_5 \, OH]^{+2} \rightleftharpoons$ $(H_2O)_4Al \overset{OH}{\underset{OH}{}} \overset{H_2O}{\underset{H_2O}{Al}} \overset{OH}{\underset{OH}{}} Al(H_2O)_4^{\,+5}$

$+ 2H \, O$

Figure 2. Types of reactions involved in the hydrolysis of Al (III).

Any suspended solids present in the water will produce an equal amount of sludge. The amount of sludge produced per turbidity unit is not as obvious; however, in many waters a correlation does exist. One correlation for a specific water is shown on Figure 3 and is a one-to-one relationship. Carbon, polymers, clay, etc., generally will produce one pound of sludge per pound of chemical addition. The sludge production for this specific supply may then be approximated by the following:

$$S = (0.44A + SS + TU + M) \, Q(8.34)$$

where   S = sludge produced, lb/day,
       Q = plant flow, mgd,
       A = alum dose, mg/l,
       SS = mg/l of suspended solids in raw water,
       TU= turbidity of filtered raw water, and
       M = mg/l of miscellaneous chemical additions such as clay, polymer and carbon.

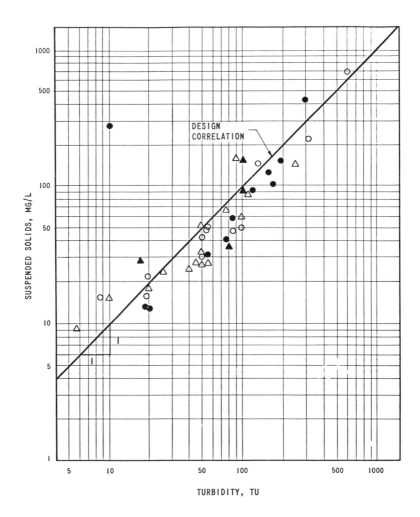

Figure 3. Suspended solids versus turbidity, Queen Lane and Belmont plants, city of Philadelphia, Pennsylvania.

Alum sludge leaving the sedimentation basin generally has a suspended solids content of 1%. Of the solids, 20-40% are volatile; the remainder are inorganic clays or silts. The BOD$_5$ of alum sludge is usually 100 mg/l. However, the chemical oxygen demand (COD) of the sludge is considerably higher. The pH of alum sludge is generally in the 5.5-7.5 range. Alum sludge from sedimentation basins may include large numbers of microorganisms, but generally does not exhibit an unpleasnat odor. The sludge flowrate is often in the range of 0.3-1% of the treatment plant flow.

Table I shows the results of specific resistance tests on sludge samples performed by other researchers. Specific resistance is used when making decisions about dewatering of sludge. Although the results shown in Table I represent widely varying conditions, they all show that alum sludge does not dewater well. Calkins and Novak [2] have stated that sludges with a specific resistance greater than 5×10^{12} m/kg (5×10^9 s^2/g) filter poorly.

Filter Backwash

All water treatment plants that practice filtration produce a large volume of washwater containing a low suspended solids concentration. The volume of backwash water is usually 2–3% of the treatment plant flow. The solids in the backwash water resemble those found in the sedimentation units. Since filters can support biological growth, the filter backwash may contain a larger fraction of volatile solids than that in the solids from the sedimentation basins. Table II shows data on the backwash characteristics from four aluminum coagulation plants.

Table I. Characteristics of Alum Sludges

Location	Water Source	Suspended Solids (%)	BOD$_5$ (mg/l)	COD (mg/l)	Specific Resistance (m/kg)
General	–	0.3-1.5	30-100	500-10,000	–
General	–	0.2-2.0	–	30- 5,000	20×10^{12}
Milwaukee. WI	Lake	0.78	–	–	5×10^{12}
Auburn, AL	–	0.19	–	–	5.5×10^{12}
Unknown	–	0.59	–	–	55×10^{12}
Rochester, NY	Lake	0.36	36-77	500- 1,000	–
San Francisco Bay, CA	Reservoir	1.0	100	2,300	–
Erie County, MI	Lake	0.16	–	–	–
Moberly, MO	Reservoir	–	–	–	16×10^{12}
Unknown	–	0.6-1.5	–	–	3.0×10^{12}
Unknown	–	–	–	–	40×10^{12}
Radford, VA	River	3.0	–	–	1.6×10^{12}
Timbersville, VA	River	1.4	–	–	5.4×10^{12}
Harrisonburg, VA	River	2.1	–	–	3.6×10^{12}
Blacksburg, VA	River	3.1	–	–	4.1×10^{12}

Table II. Characteristics of Filter Backwash Water (aluminum coagulation plants)

Plant	BOD$_5$ (mg/l)	COD (mg/l)	pH	Dry Solids, mg/l			
				Total	Volatile	Suspended	
						Total	Volatile
A	4.2	28	7.8	121	44	47	31
B	3.7	75	7.2	378	115	104	53
C	2.8	160	7.8	166	45	75	40
D	1.6	–	–	–	–	100	60

MINIMIZATION OF SLUDGE GENERATION

The minimization of sludge generation can have an advantageous effect on the requirements and economics of handling, treatment and disposal of water treatment plant sludges. Minimization also results in the conservation of natural resources (raw materials, energy and labor).

Sludges generated in water treatment usually undergo a stepwise process, which involves volume reduction by one or several thickening and dewatering processes, followed by ultimate disposal. Minimization of sludge handling costs may be achieved either by the reduction of sludge volumes or quantities, or by improving the dewatering and handling properties of the residue.

The total amount of sludge produced is the sum of the residue of the solids added during treatment, and subsequently precipitated, and the quantity of solids removed from the raw water. Both sources must be considered in deciding on methods for sludge volume reduction. Sludge produced from chemical addition in the treatment process may be reduced by using polymer coagulants in place of metal coagulants, which produce gelatinous hydroxide precipitates. This method has found application in some treatment facilities; but before polymers find widespread use there is much work to be done to determine polymer selection, required concentration and mixing requirements. The reduction of each 1 mg/l of alum will result in a savings of approximately 3,000 lb/yr of alum and reduce the sludge quantities by approximately 1,300 lb/yr for each 1 mgd of water treated.

There are four methods for minimizing the quantity of metal hydroxide precipitates in the sludge:

1. Change the water treatment process to direct filtration.
2. Substitute another coagulant (s) for alum or ferric chloride.
3. Conserve the coagulant.
4. Recover the coagulant.

Since coagulant recovery also serves to condition the sludge, it is addressed in a separate section.

Direct Filtration

Direct filtration is a treatment system in which filtration is not preceded by sedimentation. In most direct filtration applications preliminary treatment includes the rapid mixing of alum or another primary coagulant with the raw water and the addition of a filter aid immediately ahead of the filter. At some direct filtration plants contact basins also are included.

It appears that direct filtration has the greatest potential for significantly minimizing sludge handling, treatment and disposal operations. In this process, coagulant additions are reduced to a minimum and nearly all wastes are generated as filter backwash.

Direct filtration is most applicable to plants with a relatively stable high-quality raw water supply and no significant color problem. Under such conditions the process can produce dramatic economies in water treatment plant construction and operating costs, as well as minimize sludge generation.

Recent plant-scale studies on direct filtration at the Niagara County Water District's plant in Lockport, New York, illustrate the potential of this process [3]. This plant has an average capacity of 12 mgd based on a filtration rate of 2 gpm/sf. Raw water of extremely good quality is taken from the Niagara River. Data on the raw water quality as measured by turbidimeter in nephelometric units (TU) for a three-year period are shown in Table III.

To evaluate direct filtration, an existing filter (test filter) was modified by replacing the existing sand media in the filter with reverse-graded mixed media. The filtration area of the test filter was limited to 350 sf by the construction of a baffle in the filter box. This modification was necessary to evaluate filtration rates up to 6 gpm/sf because the existing piping from the filter limited the hydraulic capacity.

Table III. Niagara County Water District Turbidity Data (1973-75)

Sample Source	Turbidity		Data on Peaks	
	Average	Maximum	Turbidity Exceeded	Occurrences in 3 Years
Raw Water				
November-April	10	63	30	23
May-October	5	18	10	12
Finished Water (12 months)	0.5	2	1	24

Pretreatment for direct filtration involved the additions of a 2-mg/1 of alum to the raw water in a rapid mix tank having a detention time of 20-30 seconds and about 0.5 mg/1 of cationic polymer to the filter influent.

Another existing filter (control filter) was operated in parallel with the test filter to compare filtered water quality. The control filter was operated with the existing plant flash mixer and upflow clarifiers using an average of about 17 mg/1 of alum and 0.5 mg/1 of a cationic polymer as coagulants. Lime also was added at a rate of 12-13 mg/1 to aid in coagulation. All chemical dosages for conventional treatment were based on previous plant experience and jar tests at the neighboring city of Niagara Falls plant, which has a water intake in approximately the same location as the District's intake.

This plant-scale study indicated that direct filtration without contact basins at the Niagara County Water District's treatment plant was a viable process at filtration rates up to 6 gpm/sf using alum at the rate of 5-6 mg/1 and polymer filter aid at the rate of 0.5 mg/1. During the test program, the filtered water turbidity was always below the 1.0-TU limit set by the New York State Regulations for Public Water Supplies. In addition, with the proper alum dosage and filter conditioning, direct filtration produced a finished water that exceeded the NYSDH Tentative Performance Criteria in regard to turbidity and microscopic count. It should be possible to further improve the quality of water from a direct filtration process at the plant by using the existing clarifiers to provide contact time prior to filtration, terminating filter runs prior to the occurrence of high raw water turbidities and providing proper backwashing.

At the plant's average production rate of 12 mpd, the use of direct filtration represents a reduction in alum usage of approximately 400,000 lb/yr. Further, because lime is not required in the direct filtration process, the use of approximately 500,000 lb/yr of lime is eliminated. The polymer requirement is approximately 18,000 lb/yr. This treatment represents a reduction in sludge generation of about 100,000 lb/yr based on the reduction in coagulant usage. Using conventional treatment, the estimated sludge produced during treatment is 125 pounds of dry solids per mgd of water produced, or 630,000 lb/yr. For direct filtration, it is estimated that the sludge generation rate would decrease to about 80 pounds of dry solids per mgd of water produced, a reduction of 36%.

Alum Substitution

Sludge produced through the use of alum at a water treatment plant may be reduced and the sludge dewatering properties improved through the substitution of other chemicals for all or part of the alum.

Prior to the present concern for sludge treatment and disposal, many of the potential alum substitutes were evaluated solely on their ability to reduce water treatment plant operating costs and/or improve treated water quality. When the potential savings in sludge handling and disposal costs are considered, the benefits of partial or total substitution for alum will be enhanced for most plants.

Complete substitution for alum has been achieved by the use of iron salts, such as ferric chloride, chlorinated coppers and ferric sulfate as coagulants. Another method involves the use of a clay–polymer system in place of any metal-based coagulant.

Partial substitution for alum is obtained by reducing the alum dosage and adding a polymer or other coagulant aid. The net result can be substantial reduction in the quantity of alum used and the amount of sludge generated. New and improved coagulant aids continue to be developed. Most plants probably can benefit by a periodic review of the applicability of such aids.

As with any modification in a water treatment process, care must be taken to ensure that there will be no degradation in finished water quality or reliability of treatment. It is difficult to estimate the potential suitability of another coagulant based solely on laboratory testing. In many cases, the standard jar test is inappropriate because it does not provide information on the suitability of the treated water for filtration other than the removal of solids by settling. Often, as the use of alum is reduced, the most important characteristic of the treated water becomes the floc strength and the proper preparation of the water for filtration. These characteristics can best be evaluated on a small scale using a pilot filter, or on a controlled basis with plant-scale tests.

Coagulant Conservation

Excessive quantities of coagulants are being added at many water treatment plants. In some cases, more coagulant than may be required to obtain optimum coagulation is added "just to be safe." It is difficult to continuously determine the optimum coagulant dosage at a plant, especially with rapidly changing raw water conditions. With relatively inexpensive metal coagulants, the incentive has not existed for closer control. At some plants, the careless use of excessive quantities of coagulants has added substantially to the cost of operations, especially when considering the cost of handling, treatment and disposal of the extra solids generated by the excessive coagulant dosage.

Plant operators should be made aware of the full cost of excessive use of chemicals in water treatment. The reduction in chemical and sludge handling costs from reduced coagulant dosage is shown on Figure 4. However, while

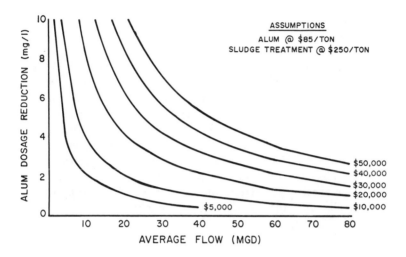

Figure 4. Reduction in chemical and sludge handling costs from reduced coagulant dosage.

from the operator's point of view an error on the high side of chemical addition is unlikely to result in an unsatisfactory product, an error on the low side may. To obtain better control, the operator probably will require additional control facilities.

At a water treatment plant in western New York State obtaining water from Lake Erie with an average daily water production of 50 mgd, the alum addition averaged 17 mg/l. Dosages were based on infrequent jar test results and the operator's experience. With the utilization of information obtained from the installation of a pilot filter unit and increased operator awareness, the average alum dosage was decreased to about 12 mg/l without deterioration in finished water quality. Alum usage was reduced by over 750,000 lb/yr, representing a savings of more than $32,000/yr and an estimated decrease in waste sludge generation of about 200,000 lb/yr of dry solids.

Softening Sludge

The key to minimizing softening sludges is to reduce the quantity of chemical addition. One should first look at how much hardness is being removed and what type. The trend in the water industry has often been to soften to 80 mg/l. However, few consumers can tell the difference at 100 mg/l and, in those states in which phosphates are still used in the detergents, plants could probably very safely soften to 140 mg/l. Each 10 mg/l of hardness left in the water reduces the sludge quantity by about 60,000 lb/yr/mgd, not an insignificant amount.

Magnesium should be removed to a final value of 40 mg/l as $CaCO_3$ hardness, as an excessive amount will cause scaling problems. However, reduction of magnesium below this level is seldom justified. The higher the magnesium hydroxide content of the sludge, the poorer its dewaterability.

Once a final hardness has been determined, the most efficient use of the chemicals should be made. Lime sludges often contain unused excess calcium hydroxide, which can be minimized by improved mixing or recirculation of sludge. A study conducted at Vandenberg Air Force Base was able to reduce the excess Ca $(OH)_2$ in the sludge from 5% to 1% by weight by sludge recirculation to the flocculator compartment. At the same time, hardness removal efficiency increased by 11%.

CHEMICAL RECOVERY

Coagulant Recovery

In the case of coagulant sludges, coagulant recovery has the benefit of reducing the quantity of sludge, as well as conditioning the sludge for subsequent treatment processes. The principle behind coagulant recovery is to remove the aluminum (or iron) from the sludge, thereby releasing the bound water and rendering the sludge easier to dewater. Figure 5 shows the effect on sludge settling of acidification to dissolve the aluminum hydroxide. Figure 6 shows the resulting improvement in sludge dewaterability.

Alum and iron recovery can be accomplished by adding acid to solubilize the metal ion salts. Sludge from settling basins is thickened. Sulfuric acid is added to the thickened sludge at approximately a stoichiometric amount as determined by the concentration of aluminum in the sludge. Generally, aluminum recovery of 60-80% can be expected by lowering the sludge pH to less than 2.5. The dissolved aluminum is separated from the residual solids by sedimentation or filtration for recycle to the plant coagulation process, while the residual sludge is disposed by landfilling after pH adjustment.

The acidic alum recovery process presents potential problems. First, the recovered alum may contain impurities, such as certain metals, which may be present in the raw water and dissolve from the sludge. Also, the recovered alum is very dilute, presenting storage and operational problems.

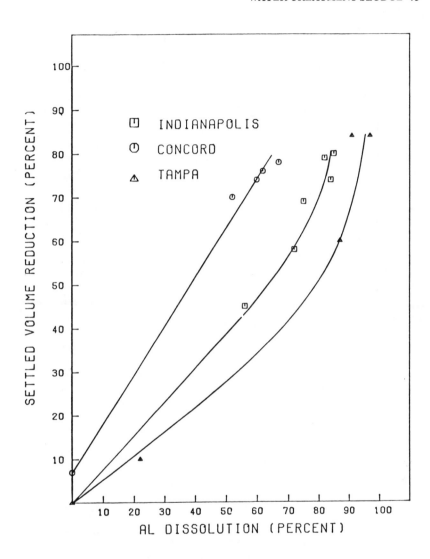

Figure 5. Effect of aluminum dissolution on sludge sedimentation.

An alternative method of alum recovery by liquid-ion exchange has been developed and patented by Michigan State University (MSU) with financial and technological support from the American Water Works Association (AWWA) Research Foundation. AWWA and MSU have set a nominal royalty charge for use of the process, with royalties going to further process development and other research projects.

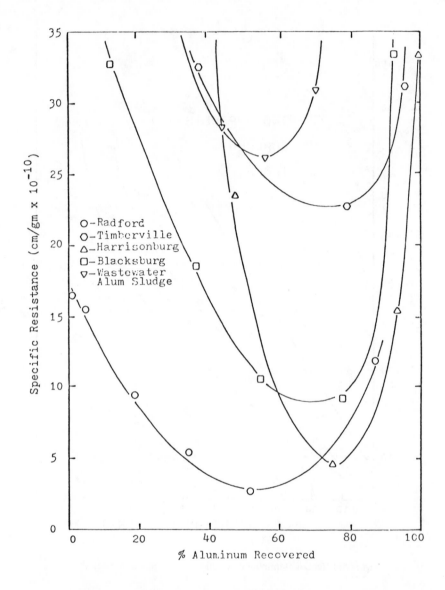

Figure 6. Variation of specific resistance of remaining sludge as a function of the percentage of aluminum recovered.

The basic liquid–ion exchange process for recovering alum from aluminum hydroxide sludge is shown schematically in Figure 7. In liquid–ion exchange, the sludge is contacted with a solvent, which removes the aluminum from the sludge. The aluminum-rich solvent is then contacted with sulfuric acid to

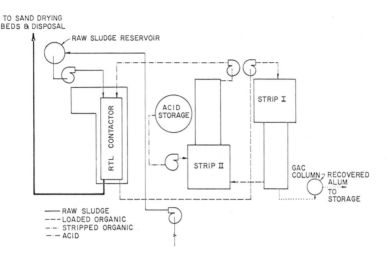

Figure 7. Schematic of liquid-ion exchange process pilot plant.

remove the aluminum from the solvent and produce alum. The same process appears to be applicable for iron coagulants.

The process has been demonstrated at pilot plants in Tampa, Florida and Niagara County, New York. Results from the process have shown:

- greater than 90% aluminum recovery;
- conditioning of the residual solids for ease in dewatering;
- production of alum of equal quality, concentration and coagulation effectiveness as commercial alum; and
- economical operation.

Lime Recovery

Lime can be recovered by recalcination. Recalcination involves separation of any impurities in the sludge from the $CaCO_3$ and then burning the $CaCO_3$ to produce CaO:

$$CaCO_3 \text{ (s)} + HEAT = CaO \text{ (s)} + CO_2 \text{ (g)}$$

The possible options are shown on Figure 8. Recalcination has these benefits.

- recovery of 1.2 times the required amount of lime;
- production of CO_2 for pH adjustment; and
- reduction in sludge weight by 80%.

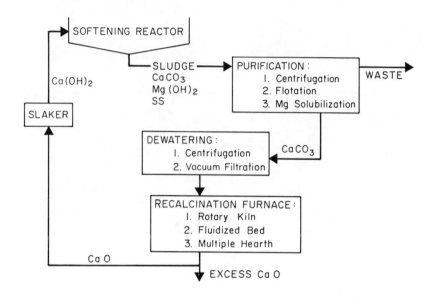

Figure 8. Options of lime recovery.

Table IV. Water Plants Practicing Lime Recovery

City	Date Installed	Production[a] of CaO (ton/day)	Type of Calciner
Miami, FL	1948	90	Rotary kiln
Salina, KS	1954	—	Flash calcination
Lansing, MI	1954	16	Fluidized reactor
Dayton, OH	1960	92	Rotary kiln
Ann Arbor, MI	1968	8	Fluidized reactor
St. Paul, MN	1968	25	Fluidized reactor
Melbourne, FL	1978	16	Multiple hearth furnace

[a] Average actual daily production.

Recalcination is not a new process and has been used in several plants, as shown in Table IV. The present high cost of energy has a significant adverse effect on the economies of recalcination.

TREATMENT ALTERNATIVES

While many methods exist for treatment and disposal of alum sludge, only a few are in widespread use. Each treatment method has its own particular problems and drawbacks. Operating costs and inability to dewater the alum sludge to a required solids content are two of the major problems with current alum sludge dewatering and disposal techniques.

Sludge is composed of solids and water. It is estimated that the percentage by weight of these elements to produce a handleable sludge cake should be at least 35% solids and 65% water. Exact percentages will depend on the requirements imposed by the selected ultimate disposal site. The objectives of a dewatering system are to produce a sludge cake suitable for land disposal and a liquid stream suitable for recycle or discharge.

Based on actions taken by the water industry to date, sludge treatment alternatives generally can be divided into three categories:

1. Co-disposal with sewage sludge at a wastewater treatment plant
2. Nonmechanical dewatering methods
3. Mechanical dewatering methods

Each of these basic alternatives is discussed in the following sections.

CODISPOSAL

A method of achieving zero discharge as far as the water treatment plant is concerned is to discharge the sludge to a sanitary sewerage system. Site-specific conditions in the water and sewerage systems will dictate the feasibility of discharging the entire sludge stream or only a concentrated portion thereof to the sewer. These same conditions will dictate the possible requirements for flow equalization.

The potential arrangements for discharging water plant sludge to the local sanitary sewerage system include:

- direct discharge of the sludge to the sanitary sewer system;
- equalization of the sludge with controlled discharge to the sanitary sewer system; and
- primary concentration of the sludge with discharge of the concentrated sludge to the sanitary sewer system, or with transport of the concentrated sludge to the wastewater treatment plant, and combining the sludge with the sewage sludge for treatment and disposal.

The direct discharge of sludge to a sanitary sewerage system (codisposal) focuses sludge treatment and disposal operations in a single location. Direct discharge, however, requires an evaluation of:

- the capability of the collection system (pipelines and pumping facilities) to handle the flow and solids loading, which may require equalization facilities to eliminate shock loadings;
- the effect on treatment facilities and operations; and
- the capital and operating costs (including industrial cost recovery charges.

When sewage treatment facilities are nearby, a separate pipe could be constructed to discharge water treatment plant sludge directly to the sewage plant's solids stream.

Codisposal of alum sludge has been practiced with success in four large U.S. cities—Detroit, Michigan; Wilmington, Delaware; Washington, DC; and Philadelphia, Pennsylvania. Although solids loadings at the wastewater treatment plants are increased because of the alum sludge, no operating difficulties have been reported. Water plant sludge disposal for the city of Philadelphia is a good example of a successful codisposal arrangement.

The principal wastes produced at the city's Queen Lane and Belmont water treatment plants are sedimentation basin sludge, filter backwash and presedimentation sludge. At the Queen Land plant, which is a ferric chloride coagulation plant, filter backwash and sedimentation basin sludge are discharged by gravity to the city sewer system and eventually processed at the Southeast Water Pollution Control Plant. Sediment from the raw water reservoir is removed by periodic dredgings and deposited in an onsite lagoon. At the Belmont plant, which is an alum coagulation plant, filter backwash, sedimentation basin sludge and raw water reservoir sediment are all discharged by gravity to the city sewer system. These wastes are processed at the Southwest Water Pollution Control Plant.

The volumes of Queen Lane and Belmont plant sludge are 4.5 and 4.6 mgd, respectively, and represent about 4% of the total flow entering the Southeast and Southwest plants. The Queen Lane sludge (27,000 lb/day) represents about 10% of the influent suspended solids to the Southeast plant; the Belmont sludge (18,000 lb/day) represents about 7% of the influent suspended solids to the Southwest plant. These solids contribute to the operating and maintenance costs of solids handling facilities at the water pollution control plants. On a dry weight basis, Queen Lane/Belmont plant sludge represents about 18% of the total quantity of sewage sludge currently produced at the Southwest plant.

The continuation of treating the sludge from the Queen Lane and Belmont plants at the Southeast and Southwest plants was investigated as a long-term method of treating and disposing the sludge [4]. An analysis of the effects of the Queen Lane/Belmont plant sludge on the city sewer system was made and the following general conclusions drawn:

1. No operating or maintenance costs could be directly associated with transporting the water plant wastes to the water pollution control plants via the sewer system.
2. Flow surges from filter backwash discharges from the Belmont plant persist for long distances within the sewer system and may contribute to the discharges to the river, which occur during wet weather periods. If the waste flows from this plant were removed from the sewer system, overflows to the river during periods of heavy rainfall might be reduced.
3. It appears that the filter backwash discharges from the Queen Lane Plant do not create any excess overflows.
4. For both plants, the filter backwash discharges may interfere with the operation of flow-regulating and level-sensing equipment in the sewer system.

An estimate of the effects of the water plant sludge on primary sedimentation at the water pollution control plants was made based on long-tube settling tests performed on mixtures of water plant wastes and raw sewage. Additional data were obtained from analysis of the pollution control plant operating records during periods of water plant sedimentation basin drainings.

It was estimated that about 75% of the water plant sludge is removed in the primary sedimentation tanks at the water pollution control plants. Based on the test results, the peak water plant sludge discharges have a major impact on the concentration of suspended solids in the influent and effluent of the primary sedimentation tanks.

Any carryover of water plant sludge from the primary sedimentation tanks flows to secondary treatment facilities at the water pollution control plants. The literature indicates that water plant sludges have no toxic effects on the activated sludge process. If the metal coagulants (iron or aluminum) in the water plant sludge were present in soluble form, a potential for toxicity would exist. However, analyses of settled mixtures of water plant waste and sewage indicate that virtually none of the metal coagulant is present in soluble form.

It was determined that the principal impact of the water plant sludge on secondary treatment would be the increased loadings of inert suspended solids on the activated sludge and final settling processes. During peak water plant waste discharge periods it would be necessary to increase the mixed liquor suspended solids concentration (MLSS) to maintain a constant concentration of volatile suspended solids in the activated sludge process. The higher MLSS would increase the solids loadings on the final sedimentation tanks.

With uncontrolled discharge of water plant wastes to the sewer system, combined sludge production at the Southeast and Southwest plants could increase periodically by as much as 150%. By utilizing available storage at the water plants to equalize the waste solids discharge rate, the maximum increase would be reduced to about 40%. The increased sludge quantities during peak water plant waste discharge periods could affect the operation of sludge processing facilities. Additionally, the reduction of the volatile solids concentration in the sludge may affect the performance of biological processes such as digestion and composting.

Based on the evaluation of codisposal and several onsite sludge dewatering alternatives, it was found that codisposal represents the most cost-effective solution for treatment and disposal of the sludge from the city's Queen Lane and Belmont plants. To reduce the impacts of discharging the sludge on the sewer system and on the water pollution control plant operations, equalization facilities were recommended to discharge the sludge at a uniform rate over a 24-hour period.

The success of codisposal in Philadelphia and in any other large city might be attributed to the fact that the water plant sludge represents a relatively small percentage of the total solids influent to the wastewater treatment plants. However, in smaller cities, where the water plant sludge might represent a large portion of the total wastewater plant solids, some operating difficulties may be experienced, and bench scale treatability studies would be warranted to evaluate the feasibility of codisposal.

NONMECHANICAL DEWATERING

For water treatment plants, especially the smaller plants at which land is readily available at a reasonable cost, nonmechnical dewatering is frequently a suitable and advantageous alternative that can lead to the economic treatment of water treatment plant sludge. Nonmechanical dewatering can be accomplished either in open lagooons or in sand-drying beds. With the use of either of these alternatives, freeze-thawing of the sludge might be considered to improve the sludge dewaterability.

Dewatering of alum sludge using nonmechanical methods, either lagoons or sand-drying beds, has been found to occur via two basic machanisms: (1) sludge drainage—draining of the free water from the sludge; and (2) sludge drying—evaporation of water remaining after draining.

Most water plant sludges contain about 75% drainable water after settling and decanting of the supernatant. In addition, because water plant sludge generally is not easily handled after draining, an additional 10% loss of water (above that achieved through draining) via air-drying appears necessary to produce a handleable sludge.

Sludge drainage is influenced by the specific filtration resistance and the compressibility of the sludge. The specific resistance affects the rate of water loss through drainage and the compressibility affects the penetration of the sludge into the sand bed. For sludges that drain easily (low specific resistance), deep applications (several feet) may be made. For those sludges that drain poorly (high specific resistance) they must be applied at depths of less than one foot if drainage is to occur in a reasonable time.

Sludge compressibility influences sand bed performance, particularly the penetration of the sludge into the bed and the resulting plugging of the bed. Such penetration would hinder draining, require frequent sand replacement and produce a highly turbid filtrate.

Sludge drying depends for the most part on prevailing weather conditions and applied sludge depth rather than sludge characteristics. In general, the time required for air-drying after draining appears to be about three times greater than the time required for draining.

Polymer conditioning of the sludge lowers the specific filtration resistance of the sludge and increases the compressibility, thus reducing the time for draining, permitting greater applied depths and reducing the time for air-drying. Additions of between 10 and 50 mg/l of polymer generally have been found to reduce the total bed dewatering time by more than 50%.

Lagoons

Lagoons have been used extensively to thicken water plant sludges. Although lagoons perform satisfactorily in settling the sludge, there is little evidence of significant consolidation because lagoons typically are designed with impermeable bottoms, thus minimizing sludge drainage. As a result, lagoon drying generally requires large land areas and often is used as a temporary solution and/or a standby facility for a more sophisticated dewatering system.

Storage and thickening of alum sludge in lagoons has been practiced at the water treatment plants of the Monroe County Water Authority (Rochester, New York), the city of Philadelphia, Pennsylvania, the city of Baltimore, Maryland, and the Elizabethtown Water Company (Plainfield, New Jersey). At each of these facilities, average solids concentrations in the lagoons have been found to be around 10-15%. At each of these plants a more sophisticated method of dewatering the sludge has been planned.

Neubaur reported that after three years of operation, a 400-foot x 320-foot x 17-foot lagoon with a 7-foot sludge depth had an average solids concentration of only 4.3% at a loading rate of 0.37 gpd/ft^2 [5]. Solids concentration in the lagoon varied from 1.7% at the top of the sludge blanket to

14% at the bottom. King et al. [6] found that a lagoon used to hold filter backwash would remove 95 to 99% of the settleable solids and 62–90% of the suspended solids.

The lagoons at each of the above plants generally have been designed as liquid/solids separation facilities. To utilize these lagoons as a nonmechanical dewatering facility it would be necessary to modify the lagoon to include an underdrainage system. If the lagoon were modified in this manner, basin sludge could be discharged to the lagoon, decanted, then allowed to drain and air-dry to a handleable condition. Such a facility would be essentially a deep sand-drying bed or may be called a drying lagoon.

Sand-Drying Beds

The use of sand beds (similar to wastewater sludge-drying beds) has been found to be effective for dewatering water plant sludges, although their use has been somewhat limited. By contrast to a lagoon, which typically has an impermeable liner to minimize infiltration into the underlying soil, sand-drying beds utilize a sand and gravel bottom to enhance drainage of water from the sludge. A sand drying bed is designed to maximize sludge drainage and air-drying. Polymer conditioning would be provided prior to sand bed drying and residue sludge removed for ultimate disposal by a front-end loader and dump truck operation.

The filtrate from the sand-drying beds can either be recycled, treated or discharged to the watercourse, depending on quality. Laboratory testing of the filtrate should be performed in conjunction with sand-drying bed pilot testing.

Neubaur reported that solids concentrations of up to 20% were obtainable for alum sludge based on bench-scale tests [5]. Testing conditions varied with fluctuating temperatures ($69°$–$81°$). Relative humidity was 72–93% and a constant 5-mph wind was applied. Detention times varied from 70 to 100 hours. Neubaur [5] reports a loading rate for sand-drying beds of 1000 ft^3 of alum sludge applied to 2000 ft^2/day of beds, or 3.74 gpd/ft^2. He estimated the cost of this type of system to be less than $5/million gallons of processed water. Sand-drying beds usually consist of 6–9 inches of sand over an underdrain system composed of up to 12 inches of gravel and drain tiles, according to Neubaur. Sludge can be applied in up to 12-inch lifts.

Freeze Treatment

Dewatering alum sludge via either of the nonmechanical methods may be enchanced by physical conditioning of the sludge through alternate natural freezing and thawing cycles. The freeze-thaw process dehydrates the sludge particles by freezing the water that is closely associated with the particles. The freezing process takes place in two stages. The first stage reduces sludge volume by selectively freezing the water molecules. Next, the solids are dehydrated when they too become frozen. The solid mass when thawed forms granular-shaped particles. This coarse material readily settles and retains its new size and shape. The residue sludge dewaters rapidly and makes suitable landfill material.

The supernatant liquid from this process can be decanted in the lagoon, leaving the solids to dewater by natural drainage and evaporation. The addition of an underdrain will increase the dewatering characteristics of the lagoon prior to, and after, freeze-thawing. Pilot-scale lagoon systems can be utilized to evaluate this method's effectiveness and establish design parameters. If required, elimination of rain and snow from the lagoon can be accomplished by constructing a roof cover.

The potential advantages of a freeze-thaw lagoon system are as follows:

1. It is insensitive to variations in sludge quality.
2. No conditioning is required.
3. Minimum operator attention is needed.
4. It is a natural process in cold climates (winter).
5. A solids cake is more acceptable at landfills.
6. Sludge is easily worked with convential equipment.

Farrell et al. [7] reported that after natural freezing, the solids content of an alum sludge increased from 0.32% to 18%. To obtain a maximum solids concentration, freezing had to be complete. Partial freezing had little effect on improving the solids concentration of the sludge, even if repeated a number of times. Snow cover, even in extremely cold climates, also was undesirable. The researchers suggested that sludge not be put out in snow and if snow accumulated on freezing sludge, it should be removed by either melting with water or plowing off. The depth to which the sludge could be applied depended on the climate, but could range from 1 inch to 22 inches. At the lower range it was reported that the system became harder to justify economically.

Several natural freeze-thaw installations are located in New York State. At the 36-mgd alum coagulation plant of the Metropolitan Water Board of Oswego County, filter backwash is discharged to lagoons that act as decant basins. Thickened sludge is pumped from the lagoons to special freeze-thaw basins in layers about 18 inches thick. The sludge has never been deeper than 1 foot during freezing because of additional water losses. The 1-foot sludge layer becomes about 3 inches of dried material after freeze-thaw. The treated sludge has been allowed to accumulate in the basins so that ultimate disposal has not been a problem.

At the Akron (New York) Water Treatment Plant (1.5-mgd capacity), the sedimentation basins are cleaned in the spring and fall and the sludge pumped to the thickener where it is removed every three or four weeks to three drying beds. The overall dimensions for the combined beds are approximately 50 ft x 30 ft. The sludge has never been applied more than 1 foot thick, which dries to about 4 inches of solids. Sludge is removed from the drying beds during the summer and fall as it becomes dry. Some sludge that is discharged in the fall is frozen and exhibits very good dewatering and handling characteristics, like a fine sand.

Sludge freezing also may be accomplished using mechanical refrigeration, which removes the seasonal restrictions of natural freezing, but increases the power costs for the treatment plant. Wilhelm and Silverblatt found that power costs can be competitive with costs of precoat or of pretreatment chemicals required for pressure or vacuum filtration in certain situations[8]. Their study showed that freeze-treated sludge readily settled to 17–22% solids. This sludge was then further dewatered using vacuum filtration or lagoons, where a solids content of 60-70% was possible. They noted that sludge must be frozen slowly and completely between 5°F and 25°F. Applied thickness varied from 0.5-2 inches, depending on freezing time and temperature. Power costs were estimated to be 20-30% of the total cost for the system. Capital costs were estimated at 55-65%, while other operating costs were 15%.

MECHANICAL DEWATERING

The objectives of a mechanical dewatering system are to produce a sludge cake suitable for land disposal and a liquid stream suitable for recycle or discharge. Various dewatering systems have been tested on all types of water treatment plant sludges. Centrifugation, vacuum filtration, belt filtration and pressure filtration have been the most widely tested methods.

Centrifugation

Two types of centrifuges are currently used for sludge dewatering: the solid bowl and the basket bowl. For dewatering alum sludges, the solid bowl has proved to be more successful than the basket [9]. In most cases, polymers are added to condition the sludge prior to centrifugation. Centrifuges are very sensitive to changes in the concentration or composition of the sludge as well as the amount of polymer applied.

A cake dryness of 15-17% is considered a good performance for a centrifuge on alum sludge. Albrecht [10] reported solids concentrations of 15-17% with feed solids between 0.4 and 6%. Loading rates varied depending on the size of the centrifuge, but Albrecht noted that the best performance was obtained when the feedrate was 75-85% of the machine's total solids or hydraulic capacity.

Solids concentrations of 6-12% have been reported by Neubaur [5]. These solids concentrations are too low to be handled by conventional earthmoving equipment or to be landfilled. Disadvantages of centrifuges include:

- low final solids concentration,
- high power and maintenance costs, and
- sensitivity to changes in feed solids content.

Some advantages are:

- small space requirements,
- potential for process automation, and
- the ability to handle thickened or dilute sludge.

Lime-softening sludge dewaters with relative ease using a centrifuge because of its calcium carbonate content. Softening sludge typically contains 80-85% calcium carbonate, 5-10% magnesium hydroxide and 3-5% iron or aluminum hydroxide. Dewatering characteristics of softening sludges vary with location, sometimes even changing from day to day. Recarbonation converts the gelatinous magnesium hydroxide in the sludge to magnesium bicarbonate. Performance data obtained on recarbonated lime-softening sludge are similar to those for conventional lime-softening sludge.

Vacuum Filtration

The rotary drum vacuum filter applies a vacuum to a porous medium to separate solids from sludge. Two basic types of rotary drum vacuum filters are used in water treatment: (1) the traveling medium, and (2) the precoated medium filters. The traveling medium is continuously removed from the drum, allowing it to be washed from both sides without diluting the sludge in the sludge vat. Bacause of the continuous washing, the filter medium is always clean. The precoat filter is coated with 2-3 inches of filter material, which is then shaved off in 0.005-inch increments as the drum moves.

Under good conditions and with suitable pretreatment, solids concentrations above 20% are possible with traveling-belt filters. Pretreatment consists of polymers, lime or both. Tests conducted by Gates and McDermott

[11] suggested that minimum specific resistance was obtained using cationic polymers. Dosages ranged from 1.53-4.79% by weight. Anionic and nonionic polymers also improved specific resistance at doses between 0.14 and 0.25%, respectively. Bugg et al. [12], on the other hand, reported that while all three types improved filterability, the anionic polymer was by far superior to the cationic, and slightly better than the nonionic, types at any pH above 5.0. Polymer dosages ranged up to 100 ppm. They noted that polymer doses should be based on a weight-to-weight basis, as opposed to weight per volume of sludge. They found that cationic polymers were more effective than non-ionic and anionic at any pH below 5.0.

Precoating vacuum filters with diatomaceous earth has proved to be useful when filtering alum sludge. Precoating allows for successful operation under varying sludge conditions. Loading rates and other performance and operating data for a typical belt vacuum filter and a precoat filter are given in Tables V and VI [13]. Even with the improvements in sludge solids concentration using vacuum filters, they are costly to buy and install and have a high operating cost.

Belt Filtration

Belt filtration is a relatively new method for dewatering sludge. The belt filter press operates on the principle that bending a sludge cake contained between two filter belts around a roll introduces shear and compressive forces in the cake, allowing water to work its way to the surface and out of the cake, thereby reducing the cake moisture content. The belt filter press has three operating zones along the length of the unit. These include the initial draining zone, which is analogous to the addition of a drying bed; the press zone, which involves application of pressure; and a shear zone in which shear is applied to the partially dewatered cake. Recent pilot-scale results indicate that the belt filter press is capable of dewatering sludge to a higher solids content than either a vacuum filter or centrifuge.

Pressure Filtration

Filter presses are a relatively new method of sludge dewatering in the United States. At the end of 1976, five plants were using pressure filters for sludgd dewatering and four others were planned [9]. As with most of the other sludge dewatering systems, the use of conditioners is necessary to obtain a high cake solids. Lime, fly ash and polymers are all used as conditioning agents, with lime almost always required. Lime is used to raise the pH

Table V. Performance and Operating Data Obtained by the Traveling-Belt Vacuum
Filter on Alum Sludge Dewatering [13]

Feed Concentration (%)	2-6
Flowrate (liter/m^2/hr)	0.7-1.4
(gal/ft^2/hr)	2-4
Dry Solids Yield (kg/m^2/hr)	0.15-0.25
(lb/ft^2/hr)	0.75-1.25
Cake Concentration with Polymer (%)	15-17
Cake Concentration with Lime (%)	30-40
Filtrate Solids (mg/l)	100-200
Polymer Dosage (kg/ton)	3-6
(lb/ft^2/hr)	6-12
Lime Dosage (%)	30-60
Drum Speed (rpm)	0.2-0.5
Operating Vacuum (mm Hg)	254-381
(in. Hg)	10-15

Table VI. Typical Precoat Performance Data on Alum Sludge Dewatering [13]

Feed Concentration (%)	2-6
Flowrate (liter/m^2/hr)	0.7-2.1
(gal/ft^2/hr)	2-6
Dry Solids Yield (kg/m^2/hr)	0.2-0.3
(lb/ft^2/hr)	1.0-1.5
Cake Concentration (%)	30-35
Filtrate Suspended Solids (mg/l)	10-20
Solids Recovery (%)	99+
Precoat Recovery (%)	30-35
Precoat Rate (kg/m^2/hr)	0.02-0.04
(lb/ft^2/hr)	0.1-0.2
Precoat Thickness (mm)	38.1-63.5
(in.)	1.5-2.5
Drum Speed (rpm)	0.2-0.3
Operating Vacuum (mm Hg)	127-508
(in. Hg)	5-20

to around 11 and improve filterability. The lime and sludge must be mixed
for at least 30 minutes to prevent the lime from plating out on the filter
media and inside of pipes. Sometimes both lime and polymers are added for
conditiioning. Fly ash or diatomaceous earth often are used to precoat the
filter media.

Thomas reported that with feed solids of 0.5-1.5%, alum sludge cake
solids were as high as 35-40% [14]. Conditioning consisted of 1 mg/l of

polyelectrolyte, or 10% lime. Westerhoff and Gruninger [15] reported solids concentrations of up to 50% after pressure filtration. Problems with pressure filters are as follows:

1. They are a batch system necessitating sludge storage.
2. They have a high labor cost associated with them.
3. When lime is used as a conditioner, filtrate disposal causes problems due to its high (around 11) pH and significant amounts of soluble aluminum present because of high pH.

Pressure filters have been used successfully to dewater alum sludge at several locations throughout the U.S. Two such locations are in Erie County, New York, at the Erie County Water Authority's Sturgeon Point Plant, and in Rochester, New York, at the Monroe County Water Authority's Shoremont Plant [16]. Both plants include gravity settling and chemical conditioning of the sludge followed by mechanical dewatering via pressure filtration. A typical process flow diagram of the sludge treatment system used at both plants is shown on Figure 9.

Primary thickening and conditioning of alum sludge at each plant is accomplished by liquid-solids separation in conjunction with chemical additions. The primary thickening equipment includes gravity sludge thickener, a sludge decant tank and appurtenant piping and pumping facilities. Chemical conditioning facilities consist of a reaction mixer, conditioned sludge retention tank, chemical preparation and storage units, and associated piping and pumping equipment.

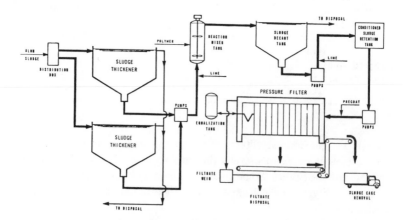

Figure 9. Alum sludge treatment process flow schematic.

At both the Sturgeon Point and Monroe County plants, mechanical dewatering is accomplished with a recessed plate-type pressure filter. The Sturgeon Point plant includes a 64-inch-diameter press, while the Monroe County facilities includes a 2-meter x 2-meter press, the largest in the U.S. for dewatering water treatment plant sludge. Other equipment associated with the pressure filtration process includes conditioned sludge equalization and pumping equipment, a precoat system, a hydraulic power system for maintaining pressure filter closure, a filtrate flow monitoring unit, a sludge bunker and sludge conveyors. Under actual operating conditions the dewatered sludge cake produced at Sturgeon Point and Monroe County contain 45-50% of dry solids by weight.

The Monroe County facility is a unique project because it treats alum sludge from three water treatment plants:

- Monroe County Water Authority's Shoremont Plant
- City of Rochester's Dewey Avenue Plant
- Eastman Kodak Company's Lake Plant

In 1971 a joint venture study was sponsored by the Monroe County Division of Pure Waters and financed by the Monroe County Water Authority, Rochester, and Kodak to investigate possible solutions to the sludge disposal problems of the three water treatment plants. Two basic approaches to solving the problem were investigated:

1. Each plant could solve its problem individually.
2. Two or three plants could cooperatively solve the problem jointly

The regional approach was selected based on the study, which included economic evaluations of the two basic alternatives. Significant capital and cost savings were indicated using the regional approach.

ULTIMATE DISPOSAL

Dewatering methods are employed to render the sludge acceptable for ultimate disposal. Attempts to use coagulant sludges as soil conditioners or stabilizers have had little success. As a result, ultimate disposal of alum sludge solids is usually accomplished by landfilling. If the sludge is to be landfilled with other material (e.g., solid waste), a solids content of above 20% is necessary at which some sludges may be handled with normal earth-moving equipment. At this solids concentration, alum sludge has the consistency of soft clay and is difficult to handle. To landfill the sludge by itself, a solids content above 40% is desirable. A solids content of 40-50% gives alum sludge the consistency of stiff clay.

Methods of sludge disposal must consider the potential pollutants in the sludge. Table VII shows the metals analyses for alum sludges from Tampa, Florida, and Norwalk, Connecticut. Landfills are often acidic because of the increased presence of carbon dioxide and the production of volatile acids from organic decomposition. The acidic condition may result in the leaching of metals from the sludge.

Technically, alum sludges are classified as industrial wastes and fall within the description of hazardous wastes. The effect that this may have on alum sludge disposal is still unclear.

In 1975 a lysimeter study for the Monroe County Water Authority in New York State evaluated the characteristics of leachate resulting from land disposal of dewatered alum sludge [17]. The effect on the quality of water passing over or through the landfill into adjacent surface waters or groundwater was determined. The study was performed using a representative alum sludge sample from the Authority's Shoremont Treatment Plant. The sludge sample was thickened, conditioned and mechanically dewatered in a pilot-scale filter press. Samples of leachate from a simulated pilot landfill containing the dewatered sludge cake were collected and analyzed for various parameters.

The data obtained from the laboratory study indicated that landfilling of

Table VII. Metals Content of Alum Sludges

Aluminum	170,000	296,296
Barium	$<200^b$	$\_^c$
Cadmium	<2	4
Chromium	>0	111
Cobolt	16	—
Copper	90	2,000
Iron	6,600	28,148
Lead	100	—
Magnesium	2,400	—
Manganese	68	9,926
Nickel	—	41
Potassium	440	—
Silver	<42	—
Sodium	2,400	—
Zinc	22	311

[a]All values are mg metal per kg of dry weight sludge.
[b]$<$ symbol indicates metal value below the indicated detection limit.
[c]— indicates data not available.

the dewatered sludge cake is a feasible practice. It was found that the higher pH of the sludge cake would help neutralize the characteristically low pH from the refuse in a sanitary landfill. It was also found that the concentrations of aluminum, chloride and iron in the leachate were well within accepted limits.

Some plants have had success in using dewatered alum sludge for landfill cover material. The use of sludge cake alone as a daily cover may result in handling problems because the cake is similar to a clay soil. The addition of sand or granular soil to the sludge cake probably would enhance its properties so that it could be used as sanitary landfill cover.

In Monroe County, New York, analyses were performed by the New York State Department of Transportation to investigate the admixing of dewatered sludge cake with soil to be used as a cover material at a sanitary landfill. The results of these analyses indicated that the sandy cover material was enhanced by mixing with sludge cake. The New York State Department of Environmental Conservation, in its review of the testing program, indicated that a landfill operation would benefit by using this mixture for daily cover. The major benefits that may be realized by landfilling dewatered sludge cake in this manner are: (1) a reduction in the cost, if any, for cover material; and (2) an increase in the capability of the cover material to fulfill certain operational functions such as minimizing moisture entering the fill, minimizing landfill odors, controlling blowing paper or dust, providing conditions for vegetative growth and minimizing erosion. Increasing the suitability of the cover material for any of these functions may be accomplished by an additional benefit of lowering the costs of providing alternative measures for performing these functions.

The proportion at which the dewatered sludge cake is mixed with existing cover material depends on the properties of the existing material. The addition of 50-100% by weight of sand or soil to the sludge cake has been suggested in previous projects to provide a suitable landfill material.

Lime sludges have been applied to farmlands as well as landfilled. A study conducted by Ohio showed that the lime required to adjust the pH of Ohio farmland into the desirable range averages 3.1 ton/ac. Maintaining the desired pH by subsequent lime applications depends on the intensity of farming. The total neutralizing power (TNP) of lime ranges from 60 to 90. Laboratory analysis of lime sludges from sewer Ohio plants showed a TNP range of 92 to 100.

Utilization of lime sludge for soil neutralization has had limited acceptance. Many farmers are dubious about using "waste" sludge for agricultural purposes to rural areas in the immediate vicinity of the water works.

Sludge can be applied to farmland by spraying liquid or thickened sludge (1-15% solids) from tank trucks or dispersing dewatered (20-40% solids) sludge from a power-driven, tank-type spreader, or from a hopper-bed truck with a spinner device for spreading the sludge.

The discharge rate of 4% sludge is 14,000 gal/ac and the application rate of 30% sludge is 8.3 ton/ac based on a requirement of 2.5 tons of lime per acre of farmland.

REFERENCES

1. Ohio Department of Health. "Waste Sludge and Filter Washwater Disposal from Water Softening Plants" (September 1969).
2. Calkins, R. J., and J. T. Novak. "Characterization of Chemical Sludges," *J. Am. Water Works Assoc.* 65:523 (1973).
3. Malcolm Pirnie, Inc. Filtration Study—Niagara County (New York) Water District, April, 1978.
4. Malcolm Pirnie, Inc. "Water Treatment Plant Waste Disposal Practices, City of Philadelphia, Pennsylvania," Volume 1 (December 1978), Volume 2 (June 1979).
5. Neubauer, A. E. "Waste Alum Sludge Treatment," *J. Am. Water Works Assoc.* 60:819 (January 1968).
6. King, P. H., et al. "Lagoon Disposal of Water Treatment Plant Wastes," *J. San. Eng. Div., ASCE* 96:103 (August 1970).
7. Farrell, B., et al. "Natural Freezing for Dewatering of Aluminum Hydroxide Sludges," *J. Am. Water Works Assoc.* 62:787 (December 1970).
8. Wilhelm, J. H., and C. E. Silberblatt. "Freeze Treatment of Alum Sludge," *J. Am. Water Works Assoc.* 66:312 (1976).
9. American Water Works Association. "Committee Report, Prepublication Report" (April 1977).
10. Albrecht, A. E. "Disposal of Alum Sludge," *J. Am. Water Works Assoc.* 64:46 (January 1972).
11. Gates, C. D., and R. F. McDermott. "Characterization and Conditioning of Water Treatment Plant Sludge," *J. Am. Water Works Assoc.* 60:331 (March 1968).
12. Bugg, H. M., P. H. King and C. W. Randall. "Polyelectrolyte Conditioning of Alum Sludges," *J. Am. Water Works Assoc.* 62:792 (December 1970).
13. Westerhoff, G. P., and M. P. Daley. "Water Treatment Plant Waste Disposal," *J. Am. Water Works Assoc.* (June 1974).
14. Thomas, C. M. "The Use of Filter Process for the Dewatering of Sludges," *J. Water Poll. Control Fed.* 43:93 (January 1971).
15. Westerhoff, G. P., and R. M. Gruninger. "Filter Plant Sludge Disposal," *Environ. Sci. Technol.* 8:122 (February 1974).
16. Dyksen, E., and R. M. Gruninger. "Success Story Times Two," *Water Wastes Eng.* (January 1979).
17. Malcolm Pirnie, Inc. Study of Dewater Alum Sludge Leachate Characteristics, February, 1976.

CHAPTER 4

IMPACT OF SEPTIC TANK SLUDGE ON MUNICIPAL WASTEWATER TREATMENT PLANTS

Raymond J. Smit, P.E., Partner and
Dr. Jon Kang, Process Specialist
McNamee, Porter and Seeley
Ann Arbor, Michigan

The U.S. Environmental Protection Agency (EPA) [1] reported in 1977 that more than 16 million onsite wastewater disposal systems were in use, serving approximately 25% of the U.S. population. Geographical distribution was reported to vary from state to state. For example, 30% of the population of Michigan is served by septic systems; in other states, the percentage of the population served by septic systems is as high as 45 percent.

With the passage of the Clean Water Act of 1977, (PL 95-217), Congress is effectively discouraging the construction of conventional 8-inch gravity sewers, while actively promoting the use of individual septic tanks, community septic tanks and other so-called "non conventional" systems. The regulations promulgated by EPA have become so difficult to satisfy that conventional gravity sewers are being eliminated from the grants program as an eligible item. The net result is that septic tanks, once considered to be a temporary solution only employed until "sewers come", are now being promoted as permanent solutions. The decrease in collector sewer construction activity will result in an increase in the numbers of individual septic tanks and an increasing need to deal with the treatment and disposal of septage.

The purpose of this chapter is to analyze the impact of septage on existing municipal treatment plants and thereby to establish proper design and operating guidelines for its handling. The scope includes the review of operating records on septage unloading, the impact analysis on each unit process and an economic analysis for treating septage. It will be shown that septage addition to a wastewater treatment plant may be a source of serious shock loadings, which generally have not been foreseen by plant designers. It will also be shown that actual costs of properly treating septage greatly exceed the charges currently being imposed by most communities.

SEPTAGE CHARACTERISTICS AND QUANTITY

Septage consists of (1) the grease and scum on the surface of the septic tank; (2) the accumulated sludge and grit at the bottom; and (3) the sewage present in the tank at the time of pumping. Septage is a highly variable anaerobic slurry having an offensive odor, the ability to foam and poor settling and dewatering characteristics.

Organic solids and nutrient concentrations are a function of the relative proportion of sludge and supernatant. These concentrations can vary by three orders of magnitude, the variability of these parameters being emphasized by the literature on septage.

Recent reports by EPA include the estimated average strength of septage, which is reproduced and shown in Table I along with the estimated concentrations used for computations.

The quantity of septage generated is dependent on many factors, such as the regulatory agency requirements on the number and size of septic tanks; local weather and soil conditions; and the average population in each dwelling unit, all of which above markedly influence the frequency of pumping. For a pumping cycle of once in three years, involving a 1000-gallon septic tank serving an average of 3.5 people, an average septage generation rate would be 95 gallons per year per capita. For the same septic tank, pumping at a four-year interval would generate 71 gallons per year per capita.

The total quantity of septage to be disposed for a given study area can be estimated by knowing the average generation rate and the population to be served within an economic distance limit.

SEPTAGE UNLOADING AT SELECTED PLANTS

An attempt was made to establish septage unloading patterns by septage haulers at selected wastewater treatment plants in Michigan. The treatment capacities of the municipal plants studied varied from 0.3–18MGD. Table II

Table 1. Septage Characteristics (mg/l)

	University of Lowel Study [2]	Washington DC Blue Plains Pilot 1973-1974 [3]	EPA Average Concentration [1]	This Chapter
Total Solids	11,600		38,800	
Total Volatile Solids	8,170		25,260	
Total Suspended Solids	9,500	13.060	13,000	10,000
Volatile Suspended Solids	7,650	8.710 (67%)	8,720	7,000
BOD$_5$	5,890	4.698	5,000	5,000
COD	19,500	23,450	42,850	20,000
Total Organic Carbon			9,930	
Total Kjeldahl-Nitrogen	410	316	677	400
Ammonia-Nitrogen	100		157	100
Total Phosphorus	190	354	253	200
Alkalinity, CaCO$_3$	610			
Grease	3,850		9,090	
pH	6.5		6.9	
LAS			157	
Metals				
Aluminum			48	
Arsenic			0.16	
Cadmium	0.1		0.71	
Chromium	0.6		1.07	
Copper	8.7		6.4	
Iron			205	
Mercury			0.28	
Manganese			5.02	
Nickel	0.4		0.9	
Lead	2		8.4	
Selenium			0.076	
Zinc	9.7		49	

Table II Septage Unloading Data

	Charlevoix	Petoskey	Coldwater	Ann Arbor
Wastewater				
Annual average flow (mgd)	0.29	0.80	1.90	18.3
Maximum month (mgd)	0.37	1.07	2.10	23.5
Design flow (mgd)	1.00	2.5	2.70	15
Operating Capacity (%)	50	50	100	100
Septage				
Yearly volume (gal)	153,000 (1979)	770,000 (1979)	623,000 (1977)	1,800,000 (1976)
Maximum monthly volume (gal)	35,200	105,000	82,000	200,000
Maximum 5-day volume (gal)	15,200	35,300	24,000	56,000
Maximum daily volume (gal)	12,500	20,800	13,000	23,000

summarizes the unloading data by daily, weekly and monthly maximums and also by total volume during the year. Figure 1 shows the monthly distribution of septage unloading from septage haulers at these plants. Note that the smaller the plant, the higher the fraction of septage in the maximum month. The fraction of monthly unloading ranged between 5 and 10% of the total yearly volume of septage received for a larger plant and ranged up to 23% for the smaller plant. For all plants the septage hauling was at a minimum in January and December.

Figures 2 and 3 present daily and hourly unloading patterns, respectively. Most of the septage loads enter the wastewater treatment plant during high sewage flows and load periods corresponding to the hours between 9 AM and 4 PM. As will be discussed later, this concurrence of septage unloading and maximum sewage flow reduces the treatment efficiency unfavorably.

The contribution of septage to the loading of the treatment plant was analyzed for the selected plants and is shown in Table III. Due to the high strength of septage, its contribution to the treatment plant loading may be very significant. Suspended solids from septage contributes to the total plant load by more than 25% in smaller plants. Whether the solids handling system is anaerobic digestion or mechanical dewatering, adequate capacity should be available before septage is accepted.

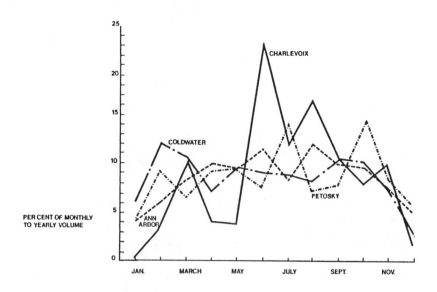

Figure 1. Monthly distribution of septage unloading.

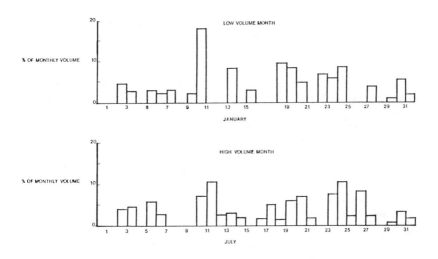

Figure 2. Daily unloading patterns in the low- and high-volume months.

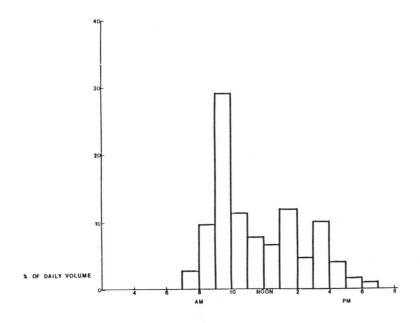

Figure 3. Hourly unloading pattern, Ann Arbor, Michigan.

Table III. Contribution of Septage to Plant Loadings

	Charlevoix	Petoskey	Coldwater	Ann Arbor
Flow				
Annual (%)	0.14	0.26	0.09	0.035
Maximum monthly (%)	0.36	0.38	0.16	0.041
Maximum 5-day (%)	1.10	1.15	–	0.068
Maximum daily (%)	4.17	2.28	0.61	0.123
TSS				
Annual average (%)	10.0	14.8	3.43	0.74
Maximum monthly (%)	28.2	24.6	7.88	2.86
BOD_5				
Maximum monthly (%)	7.3	7.9	3.8	2.0
Maximum 5-day (%)	21.1	34.0	3.8	2.6
Total Phosphorus				
Annual average (%)	4.3	11.1	2.5	0.70
Maximum monthly (%)	11.5	17.3	3.9	1.21
Ammonia Nitrogen				
Maximum monthly	–	–	–	0.18
Maximum 5-day	–	–	–	0.40

Total phosphorus in septage is also shown to be an important contributor to the plant loading. This additional phosphorus requires an increase in chemical feed to achieve phosphorus reduction required in the Great Lakes region.

In its impact on the treatment plant, the BOD_5 in septage is considered to be almost as significant as the suspended solids fraction. However, depending on the mode of operation at the plant, the impact can vary over a significant range. A question also remains to be answered as to how much of the BOD_5 or COD is associated with the settleable solids and is thus removable by primary clarification. Further studies would be helpful in clarifying this point.

There is only limited information on the impact of septage on nitrification. The amount of ammonia nitrogen appears to be small and, being very soluble, it is postulated that this fraction is washed away to the title field with each flush.

IMPACT OF SEPTAGE ADDITION ON UNIT PROCESSES

Most treatment plants that process septage introduce it into the liquid treatment process and extract the septage solids with other plant solids for processing and disposal. Some of the plants examined have tried to feed septage directly to the solids handling system; however, grit, nondecomposable solids and temperature shock loadings have been found to have an adverse effect on sludge conditioning and digestion processes. Therefore, all of the plants examined have deposited septage at the head end of the liquid flow stream.

Preliminary Treatment

Facilities for preliminary treatment typically include a receiving station, bar screen, grit removal and/or prechlorination. Control of odor has usually been poor and grit removal is often inadequate. In treatment plants, when the septage is pumped directly to digestion the grit accumulation becomes a serious problem.

A storage and equalization basin could minimize the adverse impacts of septage. The basin should be enclosed with proper odor control and provisions for periodic removal of settled grit. Where there is an existing basin for raw sewage equalization a separate basin should not be necessary. It will be wise to make provisions for direct pumping of settled sludge to a sludge handling system.

Primary Clarification

The primary clarifier can remove some settleable solids along with adsorbed materials from both the raw sewage and septage. As described earlier, however, the settling characteristics of septage vary with each load. Well-stabilized septage settles well, while septage from a failing system settles poorly.

The quantity of solids that will settle in a treatment plant varies with the mode of operation. On a continuous-feed basin, the solids will settle at normal rates. With shock loads, however, even the settleable solids will not settle well, probably because of density currents and high hydraulic velocities, both of which will create a mixing action in the clarifier. A very significant percentage of the solids may go over the weir and thus overload downstream processes. Organic materials adsorbed on solids will be carried along with solids. Primary effluents of the plants reviewed showed that BOD_5 and solids will overflow during periods of high loading.

There is a need to develop criteria for primary settling including a realistic solids unloading rate governing primary clarifier design under such loading conditions.

Activated Sludge Treatment

A mathematical model of an activated sludge plant has been generated here in an effort to better understand the impacts of septage unloading upon the plant performance. Figure 4 shows a typical flow diagram and the pertinent design parameters. A mass balance for the feed (or BOD_5) in the activated sludge process can be written as follows using simple monod kinetics:

Rate of accumulation of BOD_5 within system boundary	Rate of BOD_5 = feed into system boundary	Rate of BOD_5 − out of system boundary	Reaction of BOD_5 − oxidation within system boundary

$$(1)$$

or

$$V \cdot \frac{dc}{dt} = (QC_{PE} + Q_r C) - (Q + Q_r) C - V \cdot \frac{k\,X\,C}{K_s + C}$$

$$= QC_{PE} - QC - V \frac{k\,X\,C}{K_s + C}$$

$$\frac{dc}{dt} = \frac{Q}{V} (C_{PE} - C) - \frac{k\,X\,C}{K_s + C} \qquad (2)$$

Furthermore, both the flow and BOD_5 of the primary effluent depend on time:

$$Q = \int_q^t dt + q_0 \qquad (3)$$

$$C_{PE} = \int_0^t \frac{(dc)}{dt} dt + C_0 \qquad (4)$$

where
- V = volume of aeration tank, mg,
- Q = flowrate of primary effluent, mgd,
- Q_r = flowrate of recycle, mgd,
- q = instantaneous flowrate, mgd,
- q_0 = initial condition for q, mgd,
- C_{PE} = BOD_5 in primary effluent, mg/l,
- C = BOD_5 in secondary effluent, mg/l,
- X = MLVSS, mg/l,
- k = maximum BOD_5 removal rate per MLVSS, liter/day, and
- K_s = half-velocity constant, mg/l.

The above nonlinear differential equations were solved using numerical methods. The Continuous System Modeling Program (CSMP) developed by IBM was applied to fit the model-activated sludge operating conditions.

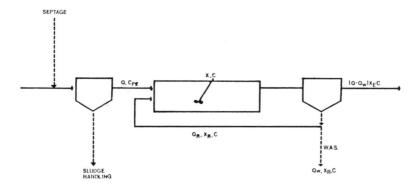

Figure 4. Schematic flow diagram of a typical activated sludge.

When the mathematical model is perturbed using normal operating conditions (typical fluctuations of both the primary flowrate and BOD_5 are as shown in Figure 5, the plant response is shown as the effluent of BOD_5. Thus, dynamic response of the system due to varying the flow and BOD_5 at each selected time interval can easily be shown.

Heavy septage unloading creates a shock load to the primary clarifier and then, in turn, to the activated sludge process. Two typical shock loads were simulated following patterns shown in Figure 3—one large shock at 10 AM and a smaller shock at 2 PM.

The dynamic responses of effluent BOD_5 predicted by the model are shown in Figure 6. The loads applied were 15% at 10 AM and 5% at 2 PM based on the daily load. The model predicts that the average BOD_5 in the effluent increased by 40% after the first shock and 60% after two shock loadings for the day compared to that of a normal day.

It should be emphasized that in the above presentation the impacts of the septage on the treatment processes were the results of shock loads. The impact of continuous feed, however, would not be as significant as the above. If the additional loads were uniformly applied to the system, the culture could become acclimated and the effluent profile would not be significantly affected.

To minimize this shock impact on effluent quality, it appears that a covered equalization basin should be provided. This basin should be sized in such a way that the secondary treatment process would receive septage at the flowrate and strength of the maximum monthly average. Any volume of septage beyond this should be stored. For the selected treatment plants, an equalization basin was sized using a mass diagram as shown in Figure 7.

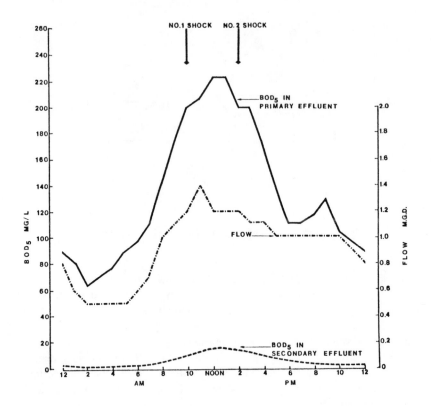

Figure 5. Typical fluctuation of flowrate and BOD$_5$ in activated sludge: vol = 0.314 mg; Q = 1 mgd; MLVSS = 1400 mg/1; K = 0.12/hr; and K$_5$ = 60 mg/1.

Note that the basin for a smaller plant should be capable of holding septage for approximately four days, while for a larger plant, approximately two days storage is suggested. Figure 8 shows the estimated detention time for equalization with respect to a total yearly volume.

Sludge Handling System

As shown in Table III, the septage contribution of solids to the plant loading reached values up to 25% of the total in small plants. The digestibility of septage solids has been reported to be very good. A typical septage has a high volatile solids content, an adequate amount of nutrients and

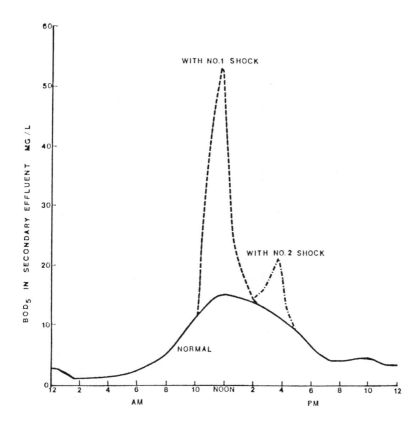

Figure 6. Dynamic response of effluent BOD$_5$ under septage shock loads.

an insignificant amount of inhibitory elements. One common problem in feeding anaerobic digesters with septage has been a sudden temperature drop when receiving an excessive amount. It appears that the septage should be introduced to the system at a slow rate following adequate flow equalization, or be pretreated to overcome temperature shocks.

Mechanical dewatering using various conditioning methods would require positive control on the percentage of septage, chemical dose and, ultimately, the sizing of all handling equipment. In summary, the sludge handling system should be sized adequately to accommodate the increased solids from septage addition.

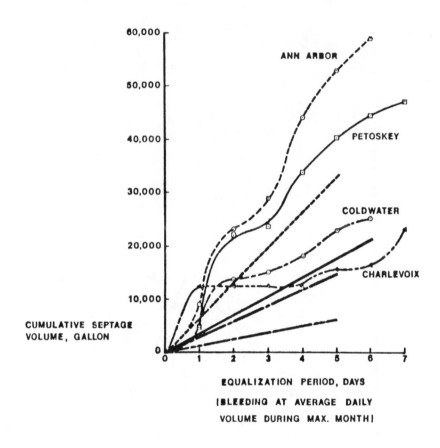

Figure 7. Graphic determination of septage flow equalization basin.

ESTIMATED TREATMENT COSTS FOR SEPTAGE

The operating budget of a treatment plant can be divided into three parts: (1) flow, (2) organic loading measured by BOD_5 and (3) suspended solids. A typical division would be 40% for flow and 30% each for BOD_5 and suspended solids. The unit costs for these parameters were estimated for various sizes of plants for secondary treatment and are shown in Table IV for capital and operation and maintenance (O&M) costs. The annual cost is based on a 25-year amortization at a 7% interest rate. The base *Engineering News Record* Index (ENR) of 3200 (1980) was used.

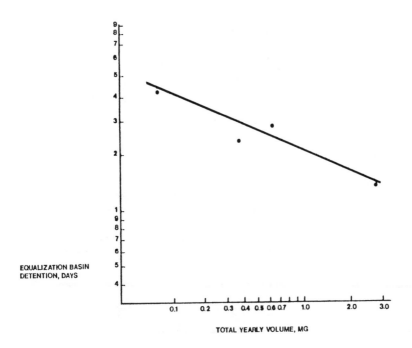

Figure 8. Septage equalization period.

Table IV. Unit Treatment Cost[a]

Design Capacity (mgd)	Capital Cost, $			O&M Cost, Dollar		
	Flow/ 1000 gal	BOD$_5$/lb	TSS/lb	Flow/ 1000 gal	BOD$_5$/lb	TSS/lb
1	0.44	0.195	0.166	0.10	0.111	0.107
2	0.38	0.168	0.143	0.85	0.094	0.091
5	0.30	0.134	0.114	0.80	0.088	0.086
10	0.25	0.110	0.093	0.63	0.070	0.067

[a]ENR = 3200; 7%, 25 year amortization.

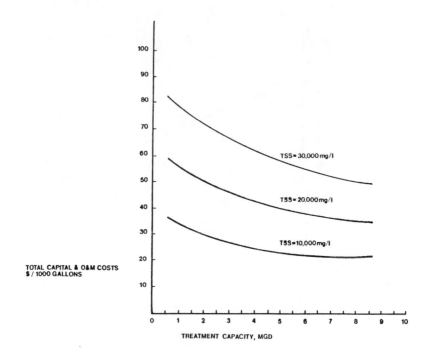

Figure 9. Total treatment cost for 1000 gallons of septage ($BOD_5 = 5000$ mg/1).

For 1000 gallons of septage, the treatment cost is estimated based on the above unit costs for capital and O&M for various sizes of treatment plants (Table V). Figure 9 also shows the treatment cost for 1000 gallons of septage having a BOD_5 concentration of 5000 mg/1 and varying concentrations of solids. It is obvious that the current rate of $5 or $10 per load of septage is far below the actual cost at the treatment plant.

It should be cautioned, however, that the above figures should not be applied universally to every treatment plant without due consideration for individual financing methods, the size of the plant, degree of treatment provided and other relevant factors.

It is important to note that the above treatment cost was computed without the participation of federal grants, which could be available under the Construction Grants Program for 75% of the capital cost. Some states also provide assistance to supplement the federal grant. When both of these grants are available, the total treatment costs could be reduced by approximately 50% from those given in Table V.

Table V. Treatment Cost for 1000 Gallons of Septage

Septage BOD$_5$ (mg/l)	Septage TSS (mg/l)	Treatment Capacity (mgd)			
		1	2	5	10
5,000	10,000	$36.00	$30.88	$26.31	$21.43
	20,000	58.78	50.40	42.98	35.02
	30,000	83.57	69.92	59.65	48.61
6,000	10,000	38.53	33.06	28.16	22.91
	20,000	61.31	52.58	44.83	36.51
	30,000	83.99	72.10	61.50	50.10
7,000	10,000	41.09	35.25	30.02	24.43
	20,000	63.87	54.77	46.69	38.02
	30,000	86.63	74.29	63.36	51.61

Land Disposal of Septage

The preceding sectons of this chapter have dealt with the combined treatment of septage with wastewater at the municipal plant. Due to the adverse impact of septage on unit processes, as discussed earlier, a separate disposal method should be explored. An economic analysis has been prepared to show a relative comparison of costs for separate septage treatment.

Land disposal of septage has merit in that septage is rich in nutrients, yet has minimal amounts of toxic or inhibitory elements for plant growth. Among several disposal techniques available only subsurface injection appears environmentally acceptable. Odor and insect problems will be well contained by the soil overburden after injection and harmful organisms will, in time, die out in the presence of the humified products. Nevertheless, a sound management policy should be followed to ensure the health and safety of the operation.

The basis of this evaluation consists of the purchase of land and a tank truck with injector, the construction for a five-month period with a mixing device of a storage tank and monitoring wells.

Table VI shows the land disposal cost for various quantities of septage. It becomes evident that separate disposal on land will not be cost-effective unless the proposed project is expected to handle a relatively large quantity of septage, such as would be possible with a county wide septage disposal program.

It is important to note again that this cost estimate did not consider financial support from federal and state grants, which may be available under

Table VI. Land Disposal Cost

Capital Cost			
Septage volume (gal/yr)	300,000	600,000	1,000,000
Land cost ($400/ac)	$ 38,400	$ 76,800	$ 129,600
Monitoring Holes	15,000	20,000	25,000
Trucks with injectors	120,000	120,000	120,000
Storage tanks	50,000	73,000	100,000
Mixers and miscellaneous	50,000	50,000	50,000
Total	$273,000	339,800	424,000
Annual Cost @ 7%, 25 years	$ 23,423	$ 29,258	$ 36,380
O&M Cost			
Truck Maintenance	$ 2,000	$ 2,000	$ 2,000
Depreciation	6,000	6,000	6,000
Fuel	1,620	3,040	$ 5,040
Labor	2,700	5,400	9,000
Sub-total	$ 12,320	$ 16,440	$ 22,040
Total Cost	$ 35,740	$ 45,700	$ 58,420
Unit Cost/1000 gallons	$ 119.13	$ 76.17	$ 58.42

the Construction Grants Program. The capital cost can be eligible for grants of up to 85% federal support plus additional state funds in some states. The total disposal cost would be reduced by approximately 50% with grant support. A final decision on separate disposal should be based on considerations of environmental safety and economics.

SUMMARY AND RECOMMENDATIONS

A careful review of records on septage unloading and its impact on the treatment process was made for several selected wastewater treatment plants in Michigan. A mathematical model was subsequently developed to predict dynamic performance of a typical activated sludge process. Finally, an economic analysis was prepared to estimate the treatment cost for septage at a municipal treatment plant. The following summarizes the conclusions and recommendations of this paper:

Septage is quite variable in its characteristics. Depending on various local conditions and regulations, the strength of septage can vary by a few orders of magnitude. In general, septage has adequate nutrients for further biological degradation and yet does not contain toxic elements.

The per capita septage generation rate varies from 71 gallons per year (for a once in four-year pumping of a 100-gallon tank) to 95 gallons per year (for a once in three-year pumping of the same tank). The frequency of pumping depends on various local conditions.

The septage unloading at the treatment plant takes place mostly during the high sewage flow and high organic loading period between 9 AM and 4 PM on weekdays. The observed pattern of unloading creates shock loads and thus affects the performance of each unit process unfavorably.

Settling characteristics of septage needs further study. In particular, interest should be focused on density currents and short circuiting characteristics produced by heavy solids loading. Design criteria could than be developed for logical clarifier loading to minimize adverse effects on subsequent unit operations.

The settleability of septage varies with each load. The fraction of BOD_5 to be removed by settling is not well known. A further study is recommended to define this variable.

A mathematical model for activated sludge was developed using Monod kinetics under varying flow and strength. The model can be used to predict the dynamic performance of an activated sludge process. The impact of shock loads on the effluent quality was illustrated with varying time and a need for septage flow equalization became quite evident. Septage equalization with adequate detention is recommended to minimize the adverse impact of shock loads. The proper detention time can be graphically determined from a mass diagram.

Economic analysis showed that the cost for treating septage is considerably more than the usual $5 or $10 per load. Charge rates should be set to cover the actual costs of construction and operation of the septage treatment system.

Separate land application of septage could be a viable alternative if the septage disposal system is planned on a regional basis.

ACKNOWLEDGMENTS

The authors of this chapter would like to acknowledge the excellent cooperation of the following officials, whose support made this chapter possible:

Mr. Lloyd Kuebler, Charlevois, Michigan
Mr. Don Steffel, Petoskey, Michigan
Mr. Dave McKay, Coldwater, Michigan
Mr. Tom Wojoski, Ann Arbor, Michigan
Mr. Dick Sayers, Ann Arbor, Michigan

Mr. Dan D'Addona of the firm of McNamee, Porter and Seeley provided necessary cost data. His sincere interest and cooperation are hereby acknowledged.

REFERENCES

1. U.S. Environmental Protection Agency. "Alternatives for Small Wastewater Treatment Systems," EPA-625/4-77-011(1977).
2. Segall, B.A., C.R. Olt, and W.B. Moeller "Monitoring Septage Addition to Wastewater Treatment Plants, prepublication report provided by Mr. S. Hathway, EPA, Cincinnati, OH.
3. U.S. Environmental Protection Agency. "Feasibility of Treating Septic Tank Waste by Activated Sludge," U.S. EPA-600/2-77-141(1977).

SECTION II

SLUDGE CHARACTERISTICS, COMPOSITION AND ANALYSIS

CHAPTER 5

SLUDGE CHARACTERISTICS

Gordon E. Jones, P.E., Vice President
    Wilkins and Wheaton Engineering Company
    Kalamazoo, Michigan

Thousands of wastewater treatment facilities across the country, have been, or are being, expanded to provide a higher degree of wastewater treatment. This, of course, translates into increased volumes of wastewater sludges, which must be handled and disposed of properly. The increase in sludge quantities is substantial, with many facilities required to provide a high degree of secondary or tertiary treatment. In addition, the requirement for phosphorus removal at publicly owned wastewater treatment works (POTW) has a significant impact on the total sludge production. These changes, coupled with more recent requirements for industrial pretreatment of wastewaters, substantially alter the characteristics of sludge for disposal. While domestic sewage sludge treated in a publicly owned treatment works is essentially organic in nature, traces of metals, minerals and other compounds invariably are also present. Sludges also contain pathogenic organisms, which survive the wastewater processes. Where industry discharges to the community systems, the potential for additional materials of concern in the resulting sludge is increased.

A survey by the U.S. Environmental Protection Agency (EPA) of publicly owned treatment facilities in the United States showed that approximately 40% of the facilities required upgrading because of their location on a water quality-limited segment on a river or stream. Additional sludge production by plants under expansion results in additional waste-activated sludge and chemical sludge production. It is the sludge characteristics from any facility that must be thoroughly reviewed prior to the evaluation and selection of a suitable disposal alternative.

SLUDGE COMPOSITION AND CHARACTERISTICS

Sludges at wastewater treatment facilities that require some means of final disposal vary on composition, depending on a number of factors:

Municipal Sludges

1. Type of community served by the system: the land use makeup of the community determines wastewater sludge characteristics, i.e., whether residential, commercial and degree of industrialization.
2. Type and degree of wastewater treatment provided: sludge characteristics vary depending on the type and degree of treatment—primary, secondary, tertiary, biological, chemical—and any facilities for separate industrial handling.
3. Type and degree of sludge conditioning provided onsite: sludge handling conditioning, and treatment greatly affect such characteristics as degree of digestion, aerobic-anaerobic, chemical conditioning, heat treatment and incineration residue.

Industrial Sludges

Industrial sludges also vary greatly, depending on a number of factors:

1. Type of industry and complexity of the processes;
2. Types of internal control for waste reduction;
3. Whether recycle or materials capture systems are utilized;
4. The degree of process manufacturing control; and
5. Type and degree of waste treatment provided.

Joint Municipal-Industrial Sludges

For many communities, the industrial contributions are great and sludges generated by these communities are unique, offering special problems. They also vary in terms of:

1. the local requirements of a Sewer Use Ordinance;

2. the extent and nature of pretreatment requirements; and
3. provisions for special municipal–industrial contracts.

Without pretreatment for the removal of trace elements, industrial users of a POTW may be a major source of such elements. Industrial users are not the only source, however, and it should not be presumed that all trace elements will be removed by a pretreatment program for industrial users.

In reviewing the above factors, it is obvious that sludge composition and characteristics vary greatly, each requiring a specific or in-depth analysis before considering final disposal solutions. In the last five to ten years, considerable testing and analysis have taken place for a variety of municipal and industrial sludges. This has been necessary in view of the high costs associated with sludge handling and disposal and concerns for the protection of the environment.

SLUDGE CHARACTERISTICS

Sludge testing and analysis have shown wide variations in the chemical and physical properties of the various sludges. The parameters generally measured in the analysis of sewage sludges depend on the options for final disposal. The characteristics generally include the following:

- Physical characteristics
- Basic chemical composition
- Concentrations of metals
- Btu value of sludge
- Mineral nutrients of the dry sludge solids
- Organic chemicals and pathogenic organisms

Physical Characteristics

The physical sludge characteristics depend on many factors discussed previously. The solids content of sludges requiring disposal ranges from less than 1% in the case of industrial and waste-activated sludges from municipal operations to industrial sludges of greater than 50% solids for certain paper mill and/or chemically conditioned or heat-treated sludges. Sludges also vary considerably with regard to their stability or biodegradability. While many sludges will decompose or break down rapidly when placed in the natural environment, other sludges are very inert and decompose very slowly. Certain sludge are totally inert and create problems in the consideration of ulti-mate disposal techniques. The stability of sludges is obviously very signifi-cant in terms of odor control and potential health hazards. Table I, taken from an EPA study [1], shows a range in physical characteristics for a num-ber of sludge types. Of particular importance in considering sludge disposal alternatives is, of course, moisture content of the sludge solids.

Chemical Characteristics

The typical chemical composition of various sludges is shown in Table II [2]. Ranges for raw sludge and digested sludges are listed. The table shows that while some constituents of the sludge decrease with digestion, most other constituents tend to be concentrated.

Metals in Sludges

Metal constituents in wastewater sludges also vary widely, as discussed previously. Table III [2,3] illustrates the tremendous variation for a large number of elements, depending on the nature of the community sampled. Further, it illustrates a very significant range in elements contained in sewage sludges from similar classifications, namely domestic, domestic and industrial, or industrial. In short, there is no such thing as an average value when con-sidering sludge disposal alternatives for a certain "type" of community, industry or sludges generated from certain processes.

As seen in the table, the lower values for strictly domestic wastewaters compare favorably. Values obtained from small Michigan municipalities are shown along with EPA—tabulated averages for suburban communities. There is a considerable discrepancy between maximum EPA—reported values, depending on the source of the data and publication. Note particu-larly the much higher observed values for chromium as sampled by Michigan communities and EPA data.

SLUDGE COMBUSTIBILITY

Another characteristic of wastewater sludges in the consideration of various alternative disposal methods is the Btu value of the sludge. These values also vary widely, depending on the characteristics of the wastewater entering the system and the treatment processes utilized. Table IV [1] shows a range of Btu values for various types of wastewater sludges, giving the variation in Btu values per pound of solids for primary sludges, primary and

Table I. Normal Quantities of Sludge Produced by Different Treatment Processes

Treatment Process	Quantity of Sludge (gal/million gal of sewage)	Moisture (%)	Specific Gravity of Sludge Solids	Specific Gravity of Sludge	Dry Solids (lb/million gal of sewage)
Primary Sedimentation					
Undigested	2,950	95	1.40	1.02	1,250
Digested in separate tanks	1,450	94	- -	1.03	750
Digested and dewatered on sand beds	- - -	60	- -	- - -	750
Digested and dewatered on vacuum filters	- - -	72.5	- -	1.00	750
Trickling Filter	745	92.5	1.33	1.025	476
Chemical Precipitation	5,120	92.5	1.93	1.03	3,300
Dewatered on vacuum filters	- - -	72.5	- -	- - -	3,300
Primary Sedimentation and Activated Sludge					
Undigested	6,900	96	- -	1.02	2,340
Undigested and dewatered on vacuum filters	1,480	80	- -	0.95	2,340
Digested in separate tanks	2,700	94	- -	1.03	1,400
Digested and dewatered on sand beds	- - -	60	- -	- - -	1,400
Digested and dewatered on vacuum filters	- - -	80	- -	0.95	1,400
Activated Sludge					
Wet sludge	19,400	98.5	1.25	1.005	2,250
Dewatered on vacuum filters	- - -	80	- -	0.95	2,250
Dried by heat dryers	- - -	4	- -	1.25	2,250
Septic Tanks (digested)	900	90	1.40	1.04	810
Imhoff Tanks (digested)	500	85	1.27	1.04	690

Table II. Typical Chemical Composition of Raw and Anaerobically Digested Sludge [2]

Item	Raw Primary Sludge Range	Typical	Digested Sludge Range	Typical
Total Dry Solids (TS), %	2.0-7.0	4.0	6.0-12.0	10.0
Volatile Solids (% of TS)	60-80	65	30-60	40.0
Grease and Fats (ether soluble, % of TS)	6.0-30.0	—[a]	5.0-20.0	—
Protein (% of TS)	20-30	25	15-20	18
Nitrogen (N, % of TS)	1.5-4.0	2.5	1.6-6.0	3.0
Phosphorus (P_2O_5, % of TS)	0.8-2.8	1.6	1.5-4.0	2.5
Potash (K_2O, % of TS)	0-1.0	0.4	0.0-3.0	1.0
Cellulose (% of TS)	8.0-15.0	10.0	8.0-15.0	10.0
Iron (not as sulfide)	2.0-4.0	2.5	3.0-8.0	4.0
Silica (SiO_2, % of TS)	15.0-20.0	—	10.0-20.0	—
pH	5.0-8.0	6.0	6.5-7.5	7.0
Alkalinity (mg/l as $CaCo_3$)	500-1,500	600	2,500-3,500	3,000
Organic Acids (mg/l as HAc)	200-2,000	500	100-600	200
Thermal Content (Btu/lb,	6,800-10,000	7,500[b]	2,700-6,800	4,000[c]
kg cal/g)	3.7-5.6	4.2[b]	1.5-3.7	2.2[c]

[a]— means data not shown in reference cited.
[b]Based on 65% volatile matter.
[c]Based on 40% volatile matter.

waste-activated sludges, and waste-activated sludge alone. It should be noted that Btu values vary considerably depending on the percentage of industrial wastewaters and their compositions.

NUTRIENT VALUES

Other sludge characteristics of particular importance when considering land application of sludge involve the mineral nutrient content. Municipal sewage sludges contain macroplant nutrients (e.g., nitrogen, phosphorus and potassium at levels about one-fifth those found in commercial fertilizers). Ranges in those nutrient levels are shown in Table V [1].

SUMMARY

In summary, it is obviously impossible to categorize wastewater sludges. Variables in the nature of the community or industry create considerable variation in the wastewater plant sludges. While sludge quantities will increase

Table III. Metals in Municipal Sludge (mg/kg) [2,3]
EPA Data

Element	Variation Michigan Municipal	EPA Data Ranges	Domestic Wastewater	Industrial and Domestic	Industrial	EPA Strictly Domestic Average
Ag		0-960	7-100	20-300	200-1,680	
As		10-50				
B		200-1,430				
Ba		0-3,000	50-400	700-1,350	2,600-6,400	
Be		—	600-1,000	<10-<100	<40-<100	
Cd	4-293	0-1,100	<10-<100	90-240	40-200	5
Co		0-800	<10-400	400-500	<40-500	
Cr	36-11,924	22-30,000	20-<400	260-2,650	1,240-2,700	50
Cu	208-2,592	45-16,030	50-200	960-2,300	1,640-4,700	250
Hg		0-89	95-700	2.6-5.0	0.6-3.0	2
Mn		100-8,800	1-11.2	500-6,100	640-6,100	
Ni	33-2,328	0-2,800	100-300	200-900	440-2,800	25
Pb	89-400	80-26,000	110-400	760-2,790	1,280-8,300	150
Sr		0-2,230	<200-<500	100-1,600	80-2,100	
Se		10-180	100-200			
V		0-2,100	<500-1,000	<200-500	1,000-2,000	
Zn	490-9,470	51-28,360	1,000-1,800	800-4,600	3,200-14,000	750

Table IV. Btu Values (lb/TS) [1]

	Ranges
Primary Sludge	8,000-13,000
Primary and Waste-Activated	5,000-10,000
Waste-Activated	5,000- 9,000

Table V. Mineral Nutrients (% of dry sludge solids) [1]

	Range
Total Nitrogen	3.5-6.4
Organic Nitrogen	2.0-4.5
Phosphorus	0.8-3.9
Potassium	0.2-0.7

in future years, it represents only a small part of the problem. Sludge characteristics such as physical, chemical and nutrient values may well be the controlling factors in choosing alternatives for ultimate sludge disposal.

REFERENCES

1. "Sludge Processing, Transportion and Disposal/Resource Recovery, A Planning Perspective," U.S. Environmental Protection Agency, Water Planning Division Report WPD 12-75-01, Washington, D.C. (1979).
2. "Municipal Sludge Management Environmental Factors," U.S. Environmental Protection Agency Office of Water Program Operations, Municipal Construction Division, EPA 430/9-77-004 Washington, DC (1977).
3. "Utilizing Municipal Sewage Wastewaters and Sludges on Land for Agricultural Production," North Central Regional Extension Publication No. 52 (1977).

CHAPTER 6

ESTABLISHING A SLUDGE SAMPLING PROGRAM

Leroy R. Dell, P.E., President
Western Michigan Environmental Services, Inc.
Holland, Michigan

Sample collection is an important part of any survey or evaluation to assess an industrial or municipal waste. Even with the most precise or accurate analytical measurements and detailed engineering calculations, the results of such a sludge evaluation survey or a process design based on that survey may be of less than desirable value without proper sample collection techniques.

The objective of a sludge sampling program is to extract a small representative sample that will accurately reflect the physical, chemical and biological characteristics of the overall sludge system at the time of sampling [1]. A number of factors determine how representative a sample can be and what degree of representativeness is actually required. These factors are reviewed for their effect on a sludge sampling program and how the program should be designed to take these factors into account.

SCOPE

The scope of this discussion is limited to presenting an outline to be followed in establishing a sampling program. Standard collection methods and procedures for particular parameters to be analyzed are set forth and well described in the literature. In particular, the American Society for Testing and Materials (ASTM) [2,3] and U.S. Environmental Protection Agency (EPA) [4-7] publications dedicate considerable text and provide detail in this area.

PURPOSE OF A SLUDGE SAMPLING PROGRAM

The purpose of a sludge sampling program must be reviewed well in advance of starting the program. This purpose will determine the extent to which the program must be carried out, as well as the required detail of the program.

One of the major requirements for a sludge sampling program today is a regulatory function. This may be to determine requirements for a permit under review or reissuance of a permit. It also may be to verify compliance with an existing permit or be in relation to legal enforcement in the case of noncompliance with or without an existing permit.

A second very important purpose is the design function. A sludge sampling program to determine design parameters for future construction can greatly affect the capital cost, operational success, and operation and maintenance (O&M) costs of a sludge disposal facility. A sludge sampling program may be established to determine the design parameters based on existing sludge. It also may be utilized to provide a data bank for design purposes from an existing process, a pilot process, a bench study or a combination of these. The sampling program may be utilized to assist in the unit process selection for a sludge disposal scheme. Results of the survey will be utilized in the design phase to provide data to determine the economics of various processes for comparison.

An additional purpose of a sludge sampling program would be as an operation function [8]. It might be utilized to determine the process settings for existing equipment. Or, in the case in which the operator has the choice of unit processes, to determine which process to utilize for the sludge being processed at that time. The sludge sampling program may be used as a bookkeeping function to determine overall efficiencies for a complete facility, for a materials balance on a process, as well as the economic effect of varying process operational parameters or sludge quality.

The sludge sampling program may be utilized in a research function to investigate other operational possibilities or the optimization of a particular process.

CONSIDERATIONS IN ESTABLISHING A
SLUDGE SAMPLING PROGRAM

After the purpose of the sludge sampling program has been evaluated effectively, the following considerations should be reviewed when establishing a sludge sampling program:

1. There are site specific conditions that must be evaluated.
2. The origin of a sludge can determine the methods by which it is sampled.
3. It is important to know the process from which the samples will be withdrawn.
4. The temperature of the sludge and the age of the sludge being assessed are important.
5. The consistency of the sludge, its flowrate and volume will have a great effect on the sampling procedures.

The facilities the sludge is contained in, or conveyed in, will determine the equipment required to properly sample the waste stream.

Hazardous materials in the sludge or conditions present in areas where the sampling must take place should be effectively evaluated. Any hazards should be assessed for their effect on the sampling personnel, the sampling procedure, preservation of the sample and the analytical procedure to be utilized in the laboratory. Another condition that must be assessed properly is seasonal or intermittent sludge production. If sampling takes place during an off season, or with a critical process offline, the results of the program will not be valid. A general layout of the facilities and the appropriate processes should be reviewed. This investigation will later assist in assessing sample points as well as evaluating overall results of the complete sludge survey.

In establishing a sludge sampling program, one area frequently overlooked is data consideration. By this is meant an assessment of the required data for the determination of sludge characteristics, which would include a review of analyses required and the degree of accuracy required by them. The desired data to assist in the assessment of the sludge should also be included, as well as additional data that may have significance later in the sludge disposal program. By including the required data, the desired data and any additional data, an evaluation can be made as to the cost of the program and how much data will be provided for that cost.

The next area considered is the actual establishment of sampling procedures. An assessment should be made as to an ideal number, sufficient volume and frequency of the samples collected. Selection of the sample point or points should take into consideration which point would be most representative of the process being sampled. This location should be reasonably accessible, safe and at a point where the flow is well mixed [9].

After picking the location, an assessment can be made of the sampling equipment [7], including the equipment required to extract the sample; the equipment for compositing multiple samples; and the equipment to provide access to sample points or for personnel protection; also included would be supplies and materials to clean the sampling equipment between samples to eliminate the possibility of any cross-contamination, and appropriate supplies to label the samples or sample containers. During the establishment of sampling procedures, a list of other pertinent observations required during the sampling should be enumerated. Appropriate preservation techniques should be assessed and listed such as type of container, the addition of preservative, refrigeration or freezing, and any onsite analysis.

Sample collection techniques should be thoroughly reviewed. In the case of a liquid sample, it should be determined whether a grab or composite sample is required. Then an assessment can be made of the appropriate technique. This decision may depend on how the sample is withdrawn. For instance, a sample could be taken from a pressure line, a pump discharge, an open tank, a closed tank, a flotation tank, multiple containers or a lagoon. Similarly, if the sample is a solid or semisolid, the type of sample (grab or composite) should be assessed. Again, the sample drawoff location is important. Examples could be a filter or other dewatering device, a contained or pressurized system, a stockpile or transport container, or an application area.

When considering composite sampling it must be assessed whether the sampling should be volume or time related, or whether multiple samples from different waste streams should be composited according to some proportion [10]. In the case of the latter, these proportions should be established prior to sampling.

The purpose of cleaning materials between samples and the method of cleaning should be evaluated early in the program. Of equal importance with previous considerations outlined is the proper labeling of samples. A multitude of unidentified samples in the laboratory would be useless.

In many cases continuous sampling is a consideration. This may involve mechanical equipment or it require frequent grab samples by personnel. Generally this would be the case where data from a sludge program would be utilized for process control.

A thorough assessment should be made of the handling of the samples following collection [11]. This should include the time required to collect and transport samples to the laboratory as well as conditions that must be maintained during that transport. A thorough review of all required records and reports should be compiled prior to starting the program so that nothing that would be of any consequence is missed at a later date. Capabilities and limitations of personnel should be reviewed to ascertain their ability to comply with the requirements of the sludge sampling program.

An evaluation of the above considerations should result in a well-conceived and outlined sludge sampling survey that will be adequate and valuable to the program requirements.

EXECUTING SAMPLE COLLECTION

With a complete sludge sampling program outlined, the next procedure is to proceed with the sample collection. All equipment, materials and supplies should be collected and reviewed. A point should be made to include extra containers as well as spare parts of expendable and breakable items. At this point, the sampling personnel should be properly trained and instructed to proceed. They should also be aware of what latitude is available to them to adjust to varying or unanticipated field conditions. Another important consideration is to correlate the sample timing with any process or weather conditions. Prior to executing sample collection, authorization to sample and obtain access to the facility should be confirmed. For example, security personnel at the facility should be aware of what is happening.

At this point it is time to proceed with the sample collection. During sampling, the containers should be protected from contamination and be labeled properly for location, date, time and preservative. This labeling should be done with a weatherproof marking system [5]. Preservation procedures should also be followed [11] and a system of monitoring the sample containers followed. Any required onsite analyses and measurements should be performed [5]. These might include such things as pH, specific conductance, temperature, chlorine residual, DO, etc. Proper care of the sampling equipment should be exercised, to avoid breakage and contamination, including thorough cleaning between samples.

Adequate record keeping while sampling must be emphasized and concisely carried out [5]. Items of major concern that should not be overlooked are weather conditions; time and date; sample or log number; description of methods and equipment used; the names of the personnel sampling and those observing; and the locations of the facility, the particular process in it and the exact sample point in this process. Additionally, the state of the operations while sampling should be noted, including flow, equipment operating, pertinent process parameters such as temperature and pressure, level, and other special operational considerations. In addition to the above, any other comments or observations should be included in the sample record report.

Following collection of the samples, proper sample transport procedures must be followed. Procedures for the chain of custody of the samples must be known, followed and recorded so that the transportation process of the samples from collection to performance of the laboratory analysis can be accounted for at a later date.

In cases where automated continuous sampling is utilized, many of the above considerations are not applicable. Generally, this type of automated sampling is for one specific purpose. A similar type of sampling that should be considered is the input to process analyzers coupled with recorders, indicators or controllers. In most cases this would be a specifically dedicated purpose and should be reviewed with the requirements of the type of the instrumentation utilized and the desired results.

FROM SAMPLING TO ANALYSIS

The principal objective when transferring a sludge sample from the collection point to the laboratory is to prevent any change in the constituents that are to be analyzed. Contamination from the container or outside sources can add to the parameters being analyzed, while losses can occur via volatilization, irreversible adsorption to container, and chemical or metabolic degradation.

Glass is the first choice for sample container material with Teflon®* being a close second. Stainless steel and aluminum are also acceptable as long as the metals involved are not to be analyzed. The EPA uses wide-mouth 16- and 32-ounce glass jars equipped with Teflon-lined caps. The Environmental Laboratory of MDNR utilizes 8-ounce wide-mouth glass jars with aluminum-lined caps.

When purgeable organics are to be analyzed for, the sample should be collected in a vial or bottle equipped with a Teflon-lined septum-sealed top. The sludge sample is placed in the bottle and organic-free water added to completely fill the container so as to leave no air space. Sample containers should be thoroughly cleaned with a detergent solution, followed by distilled water and acetone rinses. The container should then be baked in an oven at 105° C.

---

*Registered trademark of E.I. duPont de Nemours and Company, Inc., Wilmington, Delaware

No chemical preservative is needed for any of the sludge parameters. The sample should simply be iced, kept out of the sunlight and delivered to the laboratory as soon as possible after collection.

SUMMARY

In summary, to provide a proper assessment of the sludge being sampled it is necessary to exercise a well-conceived sludge sampling program. Once the above considerations have all been assessed and the conditions and equipment outlined, the resultant samples should be adequately representative of the process being sampled. It should be noted that on many occasions the cost of proper sample collection can exceed the actual cost of analysis and evaluation of the results [12]. Thus, the importance both of collecting complete and concise data during the sampling and of following prescribed procedures cannot be overemphasized.

REFERENCES

1. Sawyer and McCarty. *Chemistry for Environmental Engineering*, 3rd ed. (New York:McGraw-Hill, 1978).
2. *1979 Annual Book of ASTM Standards, Part 31, Water*, (Philadelphia, PA:American Society for Testing and Materials 1979).
3. Hamilton, C.E. *Manual on Water*, STP442A (Philadelphia, PA:American Society for Testing and Materials, 1978).
4. "Methods for Chemical Analysis of Water and Wastes," U.S. Environmental Protection Agency, Office of Research and Development (1979).
5. *NPDES Compliance Sampling Manual*, U.S. Environmental Protection Agency, Enforcement Division, MCD-51 (1979).
6. Harris and Keffer. "Wastewater Sampling Methodologies and Flow Measurement Techniques," U.S. Environmental Protection Agency, Municipal Operations Branch (1978).
7. Shelley, P.E. "Sampling of Water and Wastewater," U.S. Environmental Protection Agency, Office of Research and Development (1977).
8. Ettlich, Hinrichs and Lineck. *Operations Manual-Sludge Handling and Conditioning*, U.S. Environmental Protection Agency, Municipal Operations Branch (1978).
9. *National Handbook of Recommended Methods for Water-Data Acquisition*, U.S. Geological Survey, U.S. Department of the Interior (1977).
10. Nemerow, N.L. *Industrial Water Pollution* (Reading, PA:Addison-Wesley Publishing Co., Inc., 1978).
11. *Standard Methods for the Examination of Water and Wastewater*, 14th ed., APHA, AWWA and WPCF (1976).
12. Heidtke and Armstrong. "Probabilistic Sampling Model for Water Quality Management," *J. Water Poll. Control Fed.*, 51:2916 (1979).

CHAPTER 7

THE CONCEPTS, COSTS AND USES OF GAS CHROMATOGRAPHY AND GAS CHROMATOGRAPHY/MASS SPECTROSCOPY IN THE ANALYSIS OF SLUDGE

Richard A. Copeland, President

Environmental Research Group, Inc.
Ann Arbor, Michigan

This chapter briefly outlines the basic concepts of gas chromatography and gas chromatography/mass spectroscopy (GS/MS), and presents suggestions for waste treatment plant managers on how to keep to a minimum the organic analytical costs associated with their operating permits. The suggestions are geared to the regulatory climate in Michigan, but should be applicable to any state.

One of the most difficult problems confronting scientists, engineers and regulatory officials today is coupling the capabilities of modern analytical chemistry to the needs of environmental protection without creating a bureaucratic nightmare in terms of prohibitive cost and stifling regulations. The potential is clearly there as modern analytical techniques can now measure quantities in the parts per trillion range and regulatory agencies have been forced into taking the position that if it cannot be proved that it is not there, it must be assumed that it is.

The decade of the 1970s can be thought of as the decade of inorganic measurements. The discovery of high mercury levels in Lake St. Clair opened the decade and inorganic contaminants of all types soon were being measured everywhere. The 1980s surely will be the decade for organic analysis. The problems of polybrominated biphenyls (PBB), polychlorinated biphenyls (PCB), the Love Canal and other toxic waste dumps have been presented to the public on national television, creating a tremendous public concern for the ultimate disposal of organic wastes. Most of these wastes will wind up as industrial or municipal sludge, and most will have to be analyzed.

DESCRIPTION AND FUNCTION

Two principal instruments are used for the analysis of organic contaminants in sludge: (1) the gas chromatograph (GC), and (2) the gas chromatograph/mass spectrometer (GC/MS). Both pieces of equipment are sophisticated and very expensive, requiring highly trained technicians to operate them.

A GC separates a mixture of organic compounds into its individual components. Now separated from one another, the components, in succession, pass a detector selected to maximize the response to the class of compounds of particular interest (for example, pesticides). The detector does not identify the chemical makeup of the compound passing through it, but rather measures the amount of material passing through and the time it passes through chronologically in relation to the other compounds in the original sample.

Gas chromatography is not an exact science because it cannot positively determine the identity of a given compound; however, it can conclusively prove its absence. For example, if one were to analyze for the pesticide aldrin, one could determine by the analysis of an aldrin standard how long aldrin took to reach the detector after being injected into the gas chromatograph. By the magnitude of the detector response one has a means of quantitating output signals in relation to the quantity of aldrin injected. If a sample were introduced into the gas chromatograph and no detector response were observed at the retention time corresponding to the aldrin standard, then no aldrin is present in the sample at the detection limit for that particular gas chromatograph and detector system. If a response is observed at the exact retention time of aldrin, in most cases, it is considered to be aldrin and is quantitated and reported as such. There is no proof, however, that the compound is aldrin because it has not been shown that some other organic compound does not have the same retention time under the same conditions. For compounds in which regulatory statutes are very strict or legal liability is present, this uncertainty can be unacceptable.

The gas chromatograph/mass spectrometer overcomes this uncertainty. Instead of a simple detector such as that used on a GC, the GC/MS employs a mass spectrometer as the detector. Instead of merely measuring an electronic signal, this detector fragmentizes the compound and determines the mass (actually mass:charge ratio) and concentration of the fragments produced. This fragmentation pattern is sufficiently unique that the identity of the compound can be considered as proven for all regulatory and judicial purposes.

ANALYSIS AND ITS COST

In the course of sludge disposal it is almost certain that a chemical analysis will be required. This analysis will probably include organic compounds, which will be analyzed by the gas chromatograph or gas chromatograph/mass spectrometer. The cost of owning and operating a GC or GC/MS is very high: consequently, the cost of running analyses is high as well. With increasing pressure to require analysis for anything and everything, every effort must be made to keep analytical costs down. Due to the excessive costs of the analytical instruments, many municipalities will need to contract sludge analysis services to an outside laboratory. The following paragraphs offer suggestions for keeping these service costs to a minimum.

First, define the problem! A very common request heard by almost every analytical laboratory is the need to know what is in the sludge. The computers have access to the identity of more than 500,000 organic compounds and at least that many more exist. An analysis for a million compounds is unnecessary. Only those compounds that really need to be measured must be identified. At most, an analysis of the Michigan Critical Materials List or the U. S. Environmental Protection Agency (EPA) list of priority pollutants should be sufficient.

A second suggestion is to negotiate the analytical permit parameters with the applicable regulatory agency. Currently in Michigan, most permits are decided on a case-by-case basis. Most regulatory agencies respond conservatively; that is, when in doubt the chemical is put on the required analysis list. If there are no pesticide manufacturers within the service area concerned, there is no reason for corresponding permit to require the analysis of every legal pesticide. A specific and reasonable list specifying exactly what analyses will be required should be obtained.

Beware of the letter "s." A common permit terminology when exact chemical species are in doubt is to request analyses by classes of compounds, such as polynuclear aromatics (PNA) or phenols. Phenol (hydroxybenzene) is a single compound that can be analyzed quite inexpensively. The addition of the letter "s" to make phenols adds dozens of additional compounds to the list and increases the cost by hundreds of dollars. If the identity of specific compounds in a sludge is not initially known, a screening analysis can be performed by GC/MS and only those specific compounds detected need be on a final permit. One should not be in a position where analyses are continually requested for compounds that are never found.

Emergency service is expensive in any field and the laboratory is no exception. It is far cheaper to let the laboratory schedule the analysis within its own operating procedure than to request rush service. Plan ahead! If the results are needed in 60 days, get the samples to the laboratory well in advance. If possible, batch samples rather than submit them one at a time. Single analyses are expensive. Batch samples together with other waste treatment plants and submit them together to take advantage of quantity discounts.

SUMMARY

Nothing can be done to entirely eliminate the analytical costs associated with sludge production and disposal. However, the suggestions offered above should help most waste treatment plants keep these costs to a minimum.

CHAPTER 8

CASE HISTORY OF SAMPLING PROBLEMS
IN A SLUDGE LAGOON

John E. Schenk

Environmental Control Technology Corporation
Ann Arbor, Michigan

The ultimate disposal of treatment plant sludges is not a newly recognized problem. However, the combination of increased sludge volumes, brought about in part by more complete wastewater treatment, and the lessening of disposal options, brought about in part by an increased awareness of the polluting potential of these materials, has made this phase of wastewater management the most critical aspect for many communities and industries. Thus it has become even more critical that a comprehensive evaluation of the characteristics of these sludges be performed to fully evaluate the ultimate disposal options available. This chapter presents a case history of one such investigation into the disposal of sludges collected over a number of years in storage lagoons and serves to illustrate several details that may need to be considered in evaluating ultimate disposal options.

THE PROBLEM

For a number of years, the excess sludges from a municipal wastewater treatment plant had been stored in lagoons adjacent to the plant site. As a result of required plant expansion to provide both increased hydraulic capacity and higher treatment levels, it was established that the land occupied by the lagoons would be required for the new construction. This obviously required the removal and disposal of the approximately 40,000 cubic yards of sludge that was stored in the lagoons. The most promising method of disposal, from both a technical and economic standpoint, was landspreading. A subsequent investigation identified an apparently suitable site based on standard criteria for such disposal activities. As is so often the case, nearby homeowners were not particularly receptive to land in their immediate vicinity being used for such purposes and began to seek ways to prevent the program from being carried out. A representative of the homeowners obtained a grab sample of the sludge from the surface of the lagoon and had it analyzed for various metal ions and organic materials. Although the analytical results generally did not deviate significantly from what was anticipated, one parameter was sufficiently high to cause concern, namely polychlorinated biphenyls (PCB), at a concentration of approximately 10 mg/kg dry weight.

THE STUDY

Research has shown PCB to be potential human carcinogens. This fact, coupled with its pervasiveness due to widespread usage, has made it one of the primary environmental concerns nationwide. This is particularly true in Michigan, probably because of its alphabetic similarity to PBB. As a result, a study was initiated to define the total mass of PCB present in the sludge to be disposed of, as well as the ramifications of its presence with respect to land disposal alternatives.

The first aspect was obviously defining the total amount of PCB present in the lagooned sludge. Consequently a sampling program was carried out whereby discrete samples were obtained at depth intervals of one foot at each of five stations established on a grid network across the lagoon. Analysis of the samples obtained showed PCB concentrations in the native material underlying the sludge ranging from 0.2 to 0.4 mg/kg of dry solids (0.2–0.4 ppm), and from 2.7 to 39 mg/kg in the sludge itself. These results alone, although sufficient to show that the PCB are present in high enough concentrations to be a cause of concern, and that its concentration varied significantly throughout the lagoon, were not sufficient to fully evaluate ultimate disposal options.

To evaluate the total mass of PCB to be disposed of with the sludge, it was necessary to correlate the dry weight concentration with percent solids and density of the sludge. The density of the sludge (in kg/l) increased with depth due to compaction over the relatively long residence time. Surface samples exhibited densities of 1.08–1.10 kg/l, while samples in the 3- to

6-foot depth range exhibited densities of 1.4–1.6 kg/l. Similarly, the percent solids increased with depth, ranging from 9.0 to 45% by weight.

Using the values obtained for PCB concentration, along with percent solids and density, it was thus possible to determine the concentration of PCB on a volume basis (i.e., mg PCB/l of sludge).

RESULTS AND DISCUSSION

It is important to recognize the distinctions between the measurements on a volume versus weight basis. The volumetric concentration of a material is the information required if the total bulk of the material is going to be disposed of, such as was being considered in this case. If, however, some type of dewatering is planned to precede the disposal of the solids, it is more important to know the weight of pollutant per unit weight of solids. To illustrate the extent that these two methods of expressing the concentration of a particular material can differ, we can examine one core sample taken at this lagoon. Three separate sections were taken from this core: 0–2 feet, 2–3 feet and 3–4 feet. The respective percent solids were 13.4, 15.5 and 33.6, with densities of 1.12, 1.14 and 1.27, respectively. The measured PCB concentrations were 16, 28 and 15 mg/kg dry weight. Combining these relationships, the PCB concentrations on a volume basis were found to be 2.40, 4.95 and 6.40 mg/l, respectively. Thus, while the top and bottom samples had essentially the same weight of PCB per unit weight of solids, the differing percent solids and/or density resulted in a given volume of the bottom sludge containing nearly three times the weight of PCB as the top layer.

This volumetric information is necessary in that the requirements for land application of this sludge specified a maximum level of PCB per unit weight of soil. With this limit, along with the background levels of PCB in the native soil and the depth to which the sludge can be incorporated into the soil, it was possible to determine the total acreage required to meet established criteria. Similar evaluations will be required with respect to many organic materials and/or metal ion species potentially present in sludges.

An additional matter that often must be considered, and one of potentially critical importance in this particular case, is the potential attenuation of the material of concern after it has been spread on the land. In the case of PCB there exists the potential for biological degradation, depending on the specific type of PCB present. For example, it has been estimated that as much as 60% of mono-, di- and trichlorobiphenyls will be degraded in a one-year period. Biphenyls with four attached chloride ions may degrade as much as 20%/yr while higher substituted species will exhibit no significant degradation potential. As the objective of limiting the concentration of this material in the soil is based primarily on preventing its being present in toxic concentrations and it is assumed that proper precautions are taken to prevent the migration of this material into the air and/or groundwater, it appeared to be possible to minimize the area of land required for the spreading of sludge by determining the relative percentages of the various types of PCB present and, consequently, the relative persistance of this material over time.

SUMMARY

Based on the analysis of the sludge, it was determined that approximately 50% of the PCB was present as mono-, di- or trichlorinated, 24% present as tetrachlorinated and 26% as higher chlorinated biphenyls. With the previously mentioned assumptions for degradation rates, only 65% of the total PCB would remain after one year, 41% after three years and 34% after five years. Assuming that no other problems exist with respect to the sludge, this would indicate that only one-third of the initially required area would be necessary to provide a safe level of PCB after a five-year period.

The above example is only one illustration of the types of concerns that must be addressed to fully evaluate various sludge management alternatives. The recognition of the potentially toxic materials present in many sludges, along with constantly increasing costs for sludge management, requires that the several alternatives for sludge disposal receive enough comprehensive investigation that the most cost-effective and environmentally acceptable method is chosen.

SECTION III

LAND DISPOSAL

CHAPTER 9

AGRICULTURAL APPLICATION
OF SEWAGE SLUDGE

Lee W. Jacobs, Associate Professor

Department of Crop and Soil Sciences
Michigan State University
East Lansing, Michigan

The application of sewage sludge to agricultural land continues to be a popular choice of municipalities selecting a sludge management option. In part, this selection is due to federal legislation. Enactment of the Federal Water Pollution Control Act Amendments of 1972 (PL 92-500) initially encouraged land application as an alternative sewage treatment method. On December 27, 1977, the U. S. Congress reemphasized land application as a viable waste management alternative with the passing of the Clean Water Act of 1977. This Act provides additional financial incentives beyond those provided in PL 92-500 for innovative and alternative (I/A) approaches to waste management, which include land application of sludge.

Besides this and other legislation encouraging the recycling or reuse of waste materials, the discontinuation of ocean dumping, increased energy costs for incineration, and the difficulty encountered in locating suitable landfill sites are also determining factors. Consequently, the application of stabilized sludge to farmland competes well with other sludge management options, and the nutrient content of sewage sludge make it a good low-analysis fertilizer for crop production.

However, land application of sludge is not without its difficulties. Gaining public acceptance of this nonconventional alternative is sometimes difficult, even for small rural communities accustomed to spreading animal wastes. Land application of sludge is especially complicated for large metropolitan areas where available farmland is several miles away and frequently across political boundaries (township, county or state). It causes problems of social acceptance, high transportation costs and lack of sufficient land areas for the large quantities of sludge produced.

The purpose of this chapter is to identify the important factors that should be considered when evaluating the application of sewage sludge to agricultural soils as a sludge management alternative. An in-depth discussion of this alternative is not presented as much information is already available [1-6]. Nevertheless, the reader will find a sufficient discussion of these factors to provide a basic understanding of sludge application to farmland.

SLUDGE APPLICATION TO AGRICULTURAL LAND

Agricultural application of sewage sludge involves the spreading and incorporation of solids from wastewater treatment into the root zone of agricultural soils. Lagooning, trenching and burying of sludge in landfills are discussed in other chapters, as is high rate application of sludge to the surface of disturbed lands for reclamation purposes.

The practice of applying sludges to agricultural land should be viewed as "recycling" or "utilization" of a waste material, NOT as a "disposal" method. Using such negative terminology as disposal to discuss this sludge management option may cause people to view land application of sludge as undesirable. The use of disposal is more appropriate for other sludge management techniques discussed in other chapters.

Sludges contain nutrients and organic matter that can be utilized beneficially for growing crops. Sludges can supply appreciable amounts of nitrogen (N) and phosphorus (P) but only low quantities of potassium (K). The amount of organic matter added by an agronomic* application of sludge will be small compared to the amount of organic matter already present in most agricultural soils. However, the return of these organic solids to the soil will contribute to the maintenance of organic matter levels. Where high sludge rates are used in land reclamation programs, significant changes in soil properties can occur. Sludges stabilized by high lime treatment will also have some value as a liming material. Due to the presence of lime residues, these sludges can help neutralize acidity in soils in the same way as agricultural limestone to help maintain the proper soil pH for crop growth.

---

* "Agronomic" refers to the use of sludge in a soil–plant system at a rate to provide adequate nutrients for crop growth but not an excessive amount, which might cause pollution.

Some problems that also must be addressed for sludge application, in addition to obtaining public acceptance, are odors and aesthetics, potential pathogens in sludge, suitable site and soil, and sludge quality with respect to toxic organic chemicals, salts and heavy metals. From an agricultural perspective, the factor of greatest concern and the one most frequently encountered is high concentration of heavy metals such as cadmium, chromium, copper, lead, mercury, nickel and zinc. Since heavy metals are tied up by soils, they will accumulate in soils to the depth of incorporation when added by sludge applications. As heavy metal concentrations build up in soils, two potential problems may be encountered: (1) metal toxicity to plants, and (2) the passing along of increased amounts of metals into the food chain.

The quantity of sludge that can be applied to farmland yearly is generally based on nutrient (N and P) loadings made to supply all or part of the N and P needs of the crop to be grown. Over the long term (i.e., lifetime), the total quantity of sludge that should be applied to the soil at a land application site will be limited by the total metal loadings. The basis for determining annual and lifetime loadings of sludge to an application site is discussed in the following sections.

SLUDGE QUALITY

A necessary first step in evaluating the sludge application alternative is to determine whether a sludge is suitable for use on agricultural land. Therefore, the sludge should be analyzed to evaluate its quality. The parameters most commonly measured include percentage of total solids; total N; ammonium (NH_4^+) and nitrate (NO_3^-) nitrogen; total P and K; and total Cd, Cu, Ni, Pb and Zn. All concentrations should be on an oven-dry solids basis, using the percentage of total solids. Other elements and metals like Cr, Hg and specific organic compounds may need to be measured where industry is contributing high levels of these chemicals into the sewer system. Also, with recent federal and state legislation on toxic and hazardous waste disposal, sludges may have to be analyzed for certain hazardous chemicals in some states.

Nutrients

The nutrient concentrations are used to calculate the amount of sludge nutrients available to plants in relations to fertilizer recommendations. Not all of the N in sludges is immediately available to plants as some is present as organic N (i.e., N present in microbial cell tissue and other organic compounds). Organic N, determined by total N minus ($NH_4^+ + NO_3^-$), must be decomposed into mineral, or inorganic, forms of N like NH_4^+ and NO_3^- before plants can use this N. Therefore, the availability of sludge organic N for plants depends on the microbial breakdown in soils, as does the availability of N present in other organic materials in the soil (Figure 1).

The proportion of sludge organic N that is mineralized in a soil depends on the climate, type of soil at the application site and type of sludge applied. In the north central region, states use factors of 15–30% to estimate organic N mineralization, which occurs during the first year of crop growth. (Much lower percentages are used to estimate N mineralization occurring in the succeeding three or four years of crop growth.) Consequently, a value for the "pounds of available N per dry ton of sludge" can be calculated based on the amount of NH_4-N and NO_2-N in the sludge and the amount of organic N expected to be mineralized to NH_4^+ and No_3-N [3,7,8]. This quantity of N in the sludge will be as available in the soil for plants to use as a comparable quantity of N added by commercial fertilizers.

The total concentrations of P and K in sludge are used to calculate the additions of these nutrients as most of the P and all of the K is considered to be as available from sludge as with commercial fertilizers. To relate the

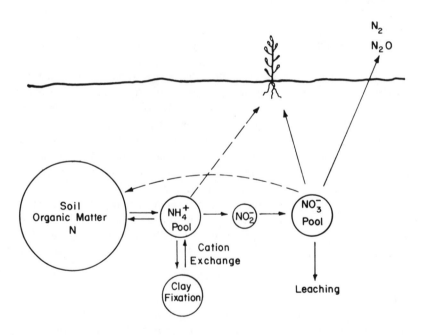

Figure 1. A simplified N cycle in soils showing how N in organic matter becomes available to plants.

calculated quantities of P and K in the sludge (i.e., "pounds of P or K per dry ton of sludge") to fertilizer recommendations normally given in pounds of phosphate (P_2O_5) and potash (K_2O), the following conversions can be used:

$$\text{Pounds P} \times 2.3 = \text{pounds } P_2O_5$$

$$\text{Pounds K} \times 1.2 = \text{pounds } K_2O$$

Sludges can vary in N and P content from a few tenths of one percent to several percent each. As indicated earlier, sludges do not contain much K, usually just a few tenths of one percent. The median concentrations of these nutrients found in sewage sludges from seven states in the north central and northeastern U. S. [3] were 3.3% N (66 lb/ton of N), 2.3% P (46 lb/ton of P or 106 lb/ton of P_2O_5), and 0.3% K (6 lb/ton of K or 7.2 lb/ton of K_2O).

Metals

Sludges can contain a wide range of heavy metal concentrations. Table I lists the range of metal concentrations found in more than 50 Michigan sludges analyzed in 1973. Also included are the maximum metal levels recommended for sludges, which would be considered good quality for application to agricultural land. When the metal concentrations in a sludge exceed these recommended levels this usually means that metals are being dis-

Table I. Ranges of Metal Concentrations Found in 57 Michigan Sludges Compared to Recommended Concentrations for Good-Quality Sludges

Metal	Range	Recommended
	ppm[a]	
Copper	84 - 10,400	1000
Nickel	12 - 2,800	200
Zinc	72 - 16,400	2500
Cadmium	2 - 1,100	25
Chromium	22 - 30,000	1000
Lead	80 - 26,000	1000
Mercury	<0.1 - 56	10

[a] ppm = parts per million and means pounds of metal per 1,000,000 pounds of dry sludge.

charged into the sewer system by industry and that more care should be exercised in using these sludges on agricultural soils. As can be seen, sludges can vary dramatically from one to another and the metal concentrations in some sludges exceed the recommended levels by more than ten times.

A major consideration regarding the use of sewage sludges on agricultural soils is the addition of heavy metals. Because metals will accumulate in soils from repeated additions, this buildup of metals can cause a plant toxicity condition in the soil or result in increased uptake of metals by plants that would pass along deleterious amounts to animals or humans. The heavy metals of greatest concern are Cd, Cu, Ni and Zn, but high concentrations of Cr, Pb, Hg and molybdenum can also be a problem [7-9]. Cadmium is of concern because of the uptake by crops and because of the adverse effects it can have on human health; whereas plant toxicity will be encountered for Cu, Ni and Zn before their concentrations in plants get high enough to adversely affect human or animal health.

The relationship between plant toxicity and the accumulation of metals is best illustrated by Figure 2. The yield curve shown relates the supply or availability of an essential plant nutrient in the soil to the amount of crop growth (or yield) [10]. Three distinct regions occur as the availability of a nutrient increases:

1. Deficiency. Supply of the nutrient is inadequate and is limiting yield, while an addition of the nutrient increases yield.
2. Sufficiency. The maximum yield has been reached and the nutrient supply is not limiting yield, so further increasing the supply in this range has no effect on yield.

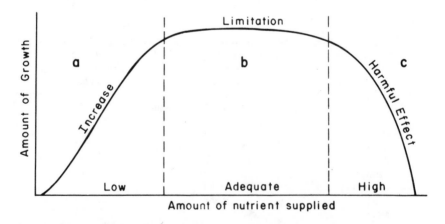

Figure 2. The relationship between growth (or yield) and the availability (or supply) of an essential plant nutrient in the soil (10).

3. Toxicity. Further additions of a nutrient beyond the sufficiency range decrease the yield until eventually no growth occurs.

This curve is applicable for Cu and Zn and other essential trace elements needed by plants, but only partly applies for Ni and other elements not essential for plant growth. For nonessential elements, increases in the available supply in soils will not affect plant growth until levels become high enough to cause toxicity.

Different guidelines have been suggested for limiting metal loadings to agricultural soils from sewage sludge applications. The approach currently having the most acceptance was proposed by a regional research committee composed of representatives from the agricultural experiment stations in north central states. These researchers recognized that the maximum safe loadings of metals will differ among soils as a consequence of differences in soil properties and in the relative toxicity of individual metals. Therefore, this committee, in cooperation with a western regional research committee, has suggested different total (i.e., lifetime) metal loadings for different agricultural soils, based on the soil's cation exchange capacity (CEC), as shown in Table II [3,7,8].

Table II. Total Amount of Metals Suggested for
Agricultural Soils Treated with Sewage Sludge

Metal	Soil Cation Exchange Capacity (meg/100g)[a]		
	<5	5-15	>15
	---Maximum amount of metal, lb/ac[b]---		
Lead	500	1000	2000
Zinc	250	500	1000
Copper	125	250	500
Nickel	50	100	200
Cadmium	5	10	20

[a] Determined by the pH 7 ammonium acetate procedure.
[b] lb/ac $\times$ 1.12 = kg/ha.

Cation exchange is an adsorption process in soils, and the number of cation exchange sites in a given weight of soil is referred to as the CEC, expressed in meg/100 g of dry soil. The CEC is a soil property that can be measured and is generally dependent on the amount and type of organic matter and clay in the soil. Even though cation exchange *is not* a principal mechanism by which metals are "tied-up" in soils, a CEC measurement reflects the clay and organic matter levels in soils. This, in turn, reflects a soil's ability to hold metals by several other mechanisms. These other mechanisms vary from one metal to another but are also generally dependent on the amount of clay and organic matter in a soil [3,7]. When metals are tied-up in soils they are not likely to cause metal toxicity to plants and soils organisms or be taken up by plants and passed along into the food chain. Thus, higher metal loadings are suggested for soils that have higher CEC values (Table II). This measurement reflects higher amounts of clay and organic matter and, therefore, a greater ability to tie-up metals.

Cadmium Limitations

Under recent U. S. Environmental Protection Agency (EPA) criteria [11], maximum cumulative Cd loadings are limited to 5 kg/ha (4.5 lb/ac), for soils with a natural background pH of less than 6.5.* For soils with a natural background pH $\geq$ 6.5 or pH $<$ 6.5 but where the pH will be maintained at or above 6.5 for as long as food-chain crops are grown, maximum cumulative Cd loadings allowed are those given in Table II, except the units for the numerical values are kg/ha instead of lb/ac (i.e., 5-10-20 kg/ha rather than 5-10-20 lb/ac).

In addition to requiring a maximum cumulative Cd loading for soils, the EPA criteria also limit annual Cd applications to soils growing food-chain crops. On land used for growing tobacco, leafy vegetables or root crops for human consumption, Cd additions cannot exceed 0.5 kg/ha/yr. Annual Cd loadings for other food-chain crops are scheduled to decrease in the future from the 2.0 kg/ha limitation currently allowed. The time schedule in the EPA criteria is:

Time Period	Annual Cd Application Rate	
	kg/ha	lb/ac
Present–June 30, 1984	2.0	1.8
July 1, 1984–December 31, 1986	1.25	1.12
Beginning January 1, 1987	0.5	0.45

---

* In semihumid climates like Michigan, the pH of soils would be expected to decrease below 6.5 due to precipitation percolating through the soil profile. Therefore, unless lime is added to maintain a more neutral soil pH in these climates, rainfall and other precipitation will cause the "natural," or unmanaged, soil pH to become acid.

Soil pH

Another factor that must be considered along with annual Cd additions and maximum cumulative loadings of the five metals listed in Table II is soil pH. The soil receiving sludge must be maintained at a pH of 6.5 or above when sludge is applied. Maintaining soil pH near neutral reaction (pH 6.5–7.0) is important because many metals are immobilized in soils at these pH levels. If soil pH is allowed to become more acid (e.g., soil pH < 6.0), metals will become more soluble in the soil and could cause metal toxicity to plants or increased metal uptake into the food chain. At the same time, a soil pH much greater than 7.0 should also be avoided because the availability of some plant nutrients is decreased.

Soluble Salts

Salts of many kinds can occur in sludges and at relatively high concentrations. High salt concentrations in the plow layer can retard or adversely affect seed germination and early growth of plant seedlings. However, soluble salts are not expected to be a problem unless sludge applications are greater than 10 dry ton/ac. Agronomic rates of sludge are usually less than this. Where higher rates are used rain will leach excess salt out of the topsoil in humid areas. Consequently, soluble salts are usually not a concern for sludge applications at agronomic rates, unless an industrial discharge is high in salt concentration.

Toxic Organics

Various types of organic contaminants may also find their way into a sewer system, depending on the kinds of wastes discharged by industry and consumers. Some of these organics will likely be broken down or decomposed during the treatment of the wastewater, but some chemicals will likely be present in sludges. The potential hazards of landspreading sludges that may contain detectable concentrations of toxic organic chemicals are largely unknown.

Recent federal and state legislation on toxic substances is requiring a cradle-to-grave approach to monitor the production, use and ultimate disposal of certain organic chemicals considered hazardous to man. Where these "critical materials" or "priority pollutants" may be present in a sewerage system, regulations will likely require that sludges produced be analyzed for these chemicals before the sludge is disposed of in a landfill or applied to land. If certain toxic organics are found at high enough concentrations to designate the sludge as a hazardous waste, application to agricultural land will likely be restricted unless the fate of those organic chemicals in soils will not pose a threat to the food chain or the environment.

SLUDGE LOADINGS TO AGRICULTURAL SOILS

In addition to the sludge analysis, other kinds of information needed for planning a sludge application program are: (1) the soil pH and lime requirement; (2) a CEC test on the soil; (3) a soil fertility chemical test; (4) the crop to be grown; and (5) the N-P-K recommendation for the crop based on the soil chemical test. The soil pH measurement and lime requirement needed to obtain a ph of 6.5 (if the soil pH is less than 6.5) are normally part of the soil chemical test. This test also assesses the availability of soil P and K. The soil fertility test results, along with crop information, are used by soil testing laboratories to make the N-P-K fertilizer recommendations. Assistance in obtaining these tests and the appropriate N-P-K recommendations can be obtained from the County Cooperative Extension Service offices or from land-grant university soil fertility extension specialists.

Finding a laboratory to do the CEC test will be more difficult than for soil fertility tests, but state extension specialists can advise on appropriate laboratories. The CEC test results are used to determine the maximum cumulative metal loadings that should be applied to the soil, as discussed earlier.

Annual Sludge Applications

Annual rates of sludge additions will be based on the N and P concentrations in the sludge and the fertilizer recommendations for the crops to be grown, or annual Cd loading limitations. Annual sludge applications at agronomic rates that satisfy crop nutrient requirements will usually be 1–5 dry ton/ac.

Sludge Nutrient Additions

For example, if one assumes that the fertilizer recommendations for a sludge application site were 150 lb/ac N, 75 lb/ac P_2O_5 and 100 lb/ac K_2O and the sludge to be applied contained 3.3% N, 2.3% P and 0.3% K, the amount of sludge needed to satisfy the N or P_2O_5 requirements can be determined. Assuming that one-third of the total N is inorganic N, each dry ton of sludge would contain about 31 lb available N, 106 lb P_2O_5 and 7 lb

K_2O. Therefore, 4.8 ton/ac (150 ÷ 31) of sludge would satisfy the N needs and 0.7 ton/ac (75 ÷ 106) of sludge would meet the P_2O_5 needs.

In many instances, using the sludge application rate based on N loadings will add more P than what is recommended. In the above example, applying 4.8 ton/ac of sludge would add more than 500 lb/ac P_2O_5, far exceeding the 75 lb/ac P_2O_5 needed. In this particular case, a lower sludge rate should be used to provide P_2O_5 and the remaining N needs supplied by supplemental N fertilizers. A rule of thumb that can be used is this: if sludge applications based on supplying all of the recommended N will add more than twice the recommended P_2O_5, a lower sludge rate should be used that will provide the needed P_2O_5.

Some additional P_2O_5 beyond what is recommended can be tolerated for a period of time. This excess P will contribute an increase in the P fertility level in a soil which can be monitored annually with soil fertility tests. However, when optimum or maximum P fertility levels are reached (as recommended by soil fertility extension specialists in your state), sludge application rates should be determined on the basis of supplying P needs.

Whether the application rate is based on supplying all the N and/or all the P needs, supplemental K_2O will be needed beyond what the sludge addition will provide. Sludges are usually low in K and agronomic rates will normally provide less than 50 lb K_2O whereas fertilizer recommendations usually suggest applying more than 50 lb/ac of K_2O. Further discussion and more examples on how to match fertilizer recommendations and sludge nutrient additions are given in the literature [3,7,8].

Annual Cd Limitation

At the current annual Cd loading allowed for agronomic crops (i.e., 1.8 lb/ac/yr), annual sludge loadings will be limited by N or P additions rather than the annual Cd limitation for most sludges. To exceed this limitation at a sludge rate of 5 ton/ac, the Cd concentrations in the sludge would have to be 180 ppm. At higher Cd concentrations, the limitation of 1.8 lb/ac/yr of Cd would further reduce the annual sludge rate below 5 ton/ac.

Cadmium concentrations at these levels are considerably higher than the 25 ppm suggested in Table I. Sludges containing more than twice this concentration of Cd should be considered poor quality and not recommended for application to agricultural land. Also, as the annual Cd limitation decreases under the time schedule discussed earlier, Cd concentrations much above 50 ppm will likely limit the annual sludge rate more frequently than N or P additions. Calculations used for determining the rate of sludge that would be needed to apply the annual Cd loading allowed are shown and discussed in the literature [3,7,8].

Physical Limitations

Limitations discussed thus far for sludge application have been based on the chemical characteristics of the sludge. Another factor to consider is the physical aspect of applying sludge to soils. The options for transporting and applying sludges to land and the costs involved are discussed adequately elsewhere [1,3]. What the reader should consider soon after calculating a rate of sludge application is what that rate means in terms of applying actual wet tons per acre.

For example, assume that the sludge used previously to calculate sludge loadings for N or P is a liquid sludge with 4.2% solids. Converting the 4.8-ton/ac and 0.7-ton/ac sludge rates calculated above to wet tons gives 114 and 16.7 wet ton/ac respectively. These rates are equivalent to about 17,400 and 4,010 liquid gal/ac, respectively. Since 1 ac-in.* of water equals approximately 27,150 gallons, about 1 inch of liquid sludge would have to be applied to apply 4.8 dry ton/ac of the assumed sludge. Physically, it would be very difficult to apply this much liquid sludge at one time or in one pass through the field.

By contrast, making a 0.7-dry ton/ac application (about 0.15 inch) would not be a problem. So, the physical problems associated with the 4.8 ton/ac rate would be another reason (besides the high P_2O_5 loadings discussed earlier) to use a lower rate of application with this sludge.

Another physical consideration in the early planning stages for a land application option is the acreage of land needed. With a given quantity of sludge, i.e., dry tons per year, almost seven times more land area will be needed if sludge were applied at 0.7 ton/ac compared to that area needed if 4.8 ton/ac were applied.

Lifetime Sludge Applications

The long-term sludge loadings for a land application site will be based on the total cumulative metal loadings made from several successive yearly applications. The total metal loadings suggested for a particular soil in some of the north central states is based on the CEC value as shown in Table II and on recent EPA criteria for Cd. Once the maximum loading is reached for any one of the five metals, sludge applications should be discontinued at that site.

---

* Volume of liquid needed to cover one acre of land to a depth of one inch.

To better understand long-term sludge loadings versus the suggested maximum metal loadings, Table III may be helpful as an illustrative example. The soil used in this example was assumed to have a CEC value of 10, and the sludge was assumed to have the metal concentrations given in column three of Table III. Each ton of dry sludge solids would contain the amount of each metal shown in the fourth column. Using these "lb/ton" concentration values for each metal, the metal additions made to the soil from an annual sludge application can be calculated by multiplying the rate of sludge applied times these values. This was done for the 4.8-ton/ac sludge rate calculated earlier when discussing N needs, and the corresponding metal loadings would be those given in column 5 of Table III. What this means is that the application of 4.8 ton/ac of this sludge would add 5.3 lb Pb, 18 lb Zn, 12 lb Cu, etc. to each acre of land.

The quantity of metals applied to a soil should be recorded each year sludge is applied so the current cumulative amount of each metal added to the soil is known relative to the total cumulative metal loadings suggested. Once the maximum loading is attained for any of the five metals, sludge applications should be discontinued. In the example shown in Table III, the maximum loading suggested for Cu, i.e., 250 lb/ac, would be reached before the maximum loadings of the other metals. This can be seen in the last column, which indicates that 104 ton/ac of the assumed sludge would be needed to apply 250 lb/ac of Cu, whereas higher quantities of sludge would be needed to apply to maximum loadings of Pb, Zn, Ni or Cd.

Table III. An Example of the Relationship Between
the Maximum Metal Loadings Suggested for a Particular Soil
and the Metals Applied by the Addition of an Assumed Sludge

Metal	Maximum Metal Loadings[a]	Concentrations of Metal in Assumed Sludge		Metal Additons at 4.8 ton/ac Sludge	Total Sludge Quantity to Apply Maximum Metal Loadings[c]
	(lb/ac)	ppm	lb/ton[b]	(lb/ac)	(ton/ac)
Pb	1000	540	1.1	5.3	909
Zn	500	1900	3.8	18	132
Cu	250	1200	2.4	12	104
Ni	100	80	0.16	0.77	625
Cd	4.5	14	0.028	0.13	161

[a] Assumed soil has CEC value of 10 and natural background soil pH $<$ 6.5. Metal loadings are from Table II except for Cd, which is limited to 5 kg/ha due to EPA criteria.
[b] Calculated by converting the concentration in ppm to pounds of metal per ton of sludge solids.
[c] Total sludge quantity is based on the metal concentrations in the sludge remaining constant with time.

More specific examples and explanations of how to calculate metal loadings, estimate lifetime utilization of a sludge application site, etc. are given in the literature [3,7,8].

LAND APPLICATION PROGRAMS

Other factors that must be considered for a successful program of applying sludges to agricultural soils include public acceptability, verbal or written agreements between farmers and municipalities, regulatory requirements, site selection, land availability and a monitoring program.

Public Acceptability

The importance of winning public acceptance of sludge application to farmland cannot be overemphasized [2,3]. A public education effort and some type of public participation program are essential for citizens to learn how this and other sludge management alternatives can be used to deal with the wastes they help to generate. During this process, the advantages and disadvantages of each alternative can be explored so that people can make an intelligent decision of their own choosing. Sludge application to land is not the best alternative for all communities; neither is any other sludge management alternative. Each community must decide what is best for its needs.

For the land application option, applying sludges to local farmland may be perceived as having adverse effects due to toxic chemicals or pathogens that may be present in sludges. Citizens may also be concerned about odors or other nuisances that could occur and the whole idea may be aesthetically displeasing to many people. The sludge may be seen by rural residents as an urban waste problem, with the urban community trying to impose its wastes on the rural community.

While technical information is available on potential health hazards [1-3, 12-16], many of the other citizen concerns are dependent on good management being utilized. A willingness of both parties to cooperate together in making the program a success and the flexibility to make changes during the initial years of the program can contribute significantly to gaining public support.

Regulatory Requirements

Early in the planning stages, information should be obtained on the local, state and federal regulations that govern sludge management alternatives. Many states have issued guidelines for the application of sludge to agricultural land and others are preparing to do so. The EPA has been developing criteria and guidelines for the past five years and longer and these have continued to change. As new knowledge about metal loadings to soils and other aspects of land application are obtained from ongoing and future research, additional changes in regulations can be expected. Therefore, the current policies of local, state and federal agencies that regulate land application should be ascertained when considering this option.

Farmer–Municipality Agreements

Verbal or written agreements between farmers and a municipality can be useful to outline how, where and when sludge can be applied. A number of arrangements have been tried, a few in which the farmer supplies part of the equipment and/or labor to apply the sludge, and some in which the farmer only provides land and the municipality handles all aspects of sludge application. In some cases a third party (contractor) is used to transport and apply sludge to farmland. Occasionally this is done at a cost to the farmer as part of an overall soil fertility package. The important point is that both parties (and sometimes a third party) must benefit economically in some way from the arrangement for the program to be successful.

Site Selection and Land Availability

Site selection procedures can begin once (1) the suitability of a sludge for application to agricultural soils has been established; (2) the public has decided to try land application for managing its sludge; and (3) a rough estimate is made to ascertain that sufficient land area is available for a land application program. A number of site selection factors must be considered [1-5]. Chief among these are the nature of the soil and its trafficability, natural and artificial drainage, permeability and water-holding capacity, and CEC. The geology and groundwater of the area, as well as land slope and potential for erosion to surface streams, may also affect site suitability.

Climate, as it relates to the length of the growing season, type of cropping pattern and accessibility to the site, is an important consideration of system design. Certain isolation distances from wells, surface streams, residences and roads may be required by regulatory agencies.

Another consideration is that farmers have land available only during certain times of the year and sludges are continuously produced. Some flexibility is possible depending on the crops grown. For example, row crops (corn, soybeans, etc.) are planted in the spring and harvested in the fall, making land available before planting and after harvesting. In comparison, cereal crops (wheat, oats, barley, etc.) are harvested in July to August, opening land much earlier for sludge application in late summer and early fall. Corn cut for silage will make land available about a month earlier in the fall than where corn is grown for grain. Nevertheless some backup contingencies are needed, such as storage or an alternative management option like landfilling or incineration, when land is not available because of a crop or wet soils.

Monitoring Program

A final factor to consider is regular monitoring of the sludge application program. Periodic sludge analyses are needed to provide nutrient and heavy metal concentrations so that rates of application can be determined to meet crop nutrient needs and total heavy metal loadings can be recorded from year to year. Annual soil chemical testing, including soil pH, will help determine N-P-K recommendations and whether any lime must be added to maintain soil pH at 6.5–7.0. How extensive a system of monitoring should be used beyond the basic testing needs will depend on the quality and amount of sludge added, the crop being grown, the level of management practices followed and other specific testing required (e.g., toxic chemicals) by regulatory agencies or the local citizenry. Complete records of sludge applications and concurrent additions of nutrients and metals made to each field should be maintained.

Monitoring should be sufficient to indicate possible problems. Where high rates of sludge are used (nonagronomic), additional monitoring of surface and tile drainage waters, groundwater, plant tissue, soils and animals may be needed. By contrast, a high degree of monitoring will not be necessary where sludge is applied to provide plant nutrients at a rate equal to fertilizer recommendations.

SUMMARY

In certain respects, designing a land application program for sludge management is no different than designing a more conventional sludge treatment system. Each can be expected to have a limited useful life for accomplishing the desired treatment of sludge and each requires a basic understanding of the processes involved. Each has its benefits and limitations.

An important consideration in designing successful land application programs is having the necessary expertise. While a certain amount of education can be accomplished by reading publications such as the ones listed, utilizing the expertise of personnel trained in agricultural soil science and/or plant science at times when uncertainties arise will be invaluable. This type of assistance may be obtained from the Cooperative Extension Services, Soil Conservation Service and consultants having this type of expertise.

Good management is the key to a successful land application program for sewage sludges. This will help ensure that potential hazards are minimized and potential benefits are maximized from a technical standpoint. At the same time, good management is critically important in achieving and keeping social acceptance by not creating public nuisances when conducting land application programs. With both management objectives in mind, many communities will find the application of municipal sewage sludge to agricultural land a good choice.

REFERENCES

1. Knezek, B. D., and R. H. Miller, Eds. "Applications of Sludges and Wastewaters on Agricultural Land: A Planning and Educational Guide," N. Central Regional Res. Pub. 235, Ohio Agricultural Resources Development Center, Wooster, OH (1976).
2. Jacobs, L. W., Ed. "Utilizing Municipal Sewage Wastewaters and Sludges on Land for Agricultural Production," N. Central Regional Extension Publ. No. 52, Cooperative Extension Services, Michigan State University, East Lansing, MI (1977).
3. Sommers, L. E. In: *Sludge Treatment and Disposal, Sludge Disposal,* Vol 2, EPA-625/4-78-012, Technology Transfer (1978), p. 57.
4. Soil Conservation Society of America. "Land Application of Waste Materials," Ankeny, IA (1976).
5. Elliot, L. F., and F. J. Stevenson, Eds. *Soils for Management of Organic Wastes and Wastewaters,* Soil Society of America, Inc., Madison, WI (1977).
6. Loehr, R. C., Ed. *Land as a Waste Management Alternative* (Ann Arbor, MI: Ann Arbor Science Publishers, Inc., 1977).

7.   Sommers, L. E., and D. W. Nelson. In: "Application of Sludges and Wastewaters on Agricultural Land: A Planning and Educational Guide," N. Central Regional Res. Pub. 235, Ohio Agricultural Resources Development Center, Wooster, OH (1976), Section 3 and App. B, p. 3.1.

8.   Galloway, H. M., and L. W. Jacobs. In: "Utilizing Municipal Sewage Wastewaters and Sludges on Land for Agricultural Production," N. Central Regional Extension Pub. No. 52, Cooperative Extension Services, Michigan State University, East Lansing, MI (1977), p. 3.

9.   Council for Agricultural Science and Technology, "Application of Sewage Sludge to Cropland: Appraisal of Potential Hazards of the Heavy Metals to Plants and Animals," Report No. 64, Council for Agricultural Science and Technology, Ames IA (1976).

10.   Russell, E. W. *Soil Conditions and Plant Growth*, 9th ed. (London: Longman Group Limited, 1961), p. 57.

11.   "Criteria for Classification of Solid Waste Disposal Facilities and Practices; Final, Interim Final and Proposed Regulations," *Federal Register* 44(179) (September 13, 1979).

12.   Burge, W. D., and P. B. Marsh, "Infectious Disease Hazards of Landspreading Sewage Wastes," *J. Environ. Qual.* 7(1):1-9 (1978).

13.   Elliott, L. F., and J. R. Ellis. "Bacterial and Viral Pathogens Associated with Land Application of Organic Wastes," *J. Environ. Qual.* 6(3): 245-251 (1977).

14.   Burge, W. D. In: *Proc. Fifth Nat. Conf. on Acceptable Sludge Disposal Techniques*, (1978), p. 125.

15.   Weaver, D. E., J. L Mang, W. A. Galke and G. J. Love. In: *Land as a Waste Management Alternative*, R. C. Loehr, Ed. (Ann Arbor, MI: Ann Arbor Science Publishers, Inc., 1977), p. 363.

16.   SCS Engineers. "Health Effects Associated with Wastewater Treatment and Disposal Systems State-of-the-Art Review," EPA-600/1-79-016a (Springfield, VA: National Technical Information Service, 1979).

CHAPTER 10

LANDFILLING AND RENOVATION
WITH SLUDGES

John C. Jenkins, Member
Jones & Henry Engineers, Ltd.
Toledo, Ohio

In the quest to improve our environment and conserve resources, land-filling and renovation are two of the sludge disposal methods considered satisfactory. One is a disposal method only; the other, an attempt to recover resources as well as dispose of sludge. This chapter presents a brief overview of these methods and a limited number of disposal operations.

Today, disposing of industrial and waste treatment sludges is complex. Higher water quality standards and improved treatment processes have led to an increased amount of sludge requiring disposal. Concurrently, environmental, regulatory, and economic concerns have limited the options available for sludge disposal.

The principal sludge disposal alternatives are incineration, composting, landspreading and landfilling. Incineration, once deemed the most practical ultimate disposal method, has been deemphasized because of high capital costs, high operating costs and intensive energy demands. Composting requires a low capital expenditure, but has high operating costs. The market for the composted product must be consistent or major backup disposal facilities must always be available.

Landspreading has been increasingly emphasized by regulatory agencies. This disposal method is particularly well suited to areas with enough of the appropriate type of land. For large urban areas, sludge must often be transported great distances for disposal, which markedly increases operating costs. Landspread sludges must be monitored carefully for metals or toxic materials. Like composting, landspreading does offer the advantage of resource recovery.

Sludge landfilling traditionally has been accepted as a viable, albeit difficult, disposal method. A properly designed and operated sludge landfill is an effective means of sludge disposal. Site selection, fill design and fill operations must be performed carefully, in strict adherence to precautions necessary for environmental protection.

Sludge can be landfilled in a codisposal landfill, where sludge is mixed with refuse or in sludge-only landfills. Sludge can also be used as a top dressing material for land reclamation projects. Each type of landfill has particular characteristics and operating techniques, although certain general design features are common to both codisposal and sludge-only landfills.

LANDFILL DESIGN CONSIDERATIONS

A landfill is a sludge or solid waste repository constructed under environmentally acceptable conditions. These conditions are generally covered in standard regulatory requirements for site selection, design and operation. Table I summarizes the design criteria established in the U.S. Environmental Protection Agency (EPA) *Process Design Manual For Municipal Sludge Landfills* [1].

Normally, a sludge landfill is constructed on agriculturally or industrially zoned land. The land should neither be archaeologically or historically significant nor environmentally sensitive. A landfill site should not be in floodplains and wetlands, and surface waters should be diverted from the site. Surface slopes should be greater than 1% to facilitate runoff, but less than 20% to avoid erosion as ordained by soil type.

Site geology should include a clay soil extending 20 feet above bedrock. The cation exchange capacity of the soil should be more than 15 meg/100g, and soil permeabilities should be less than 10^{-5} cm/sec. Local aquifers should be isolated. The gradient and quality of the groundwater system should be determined and wells installed to monitor changes in groundwater quality.

Site vegetation and access roads should be considered in site selection. Trees and shrubs can be used to block views of the operating face from nearby homes and highways. However, too much vegetation will increase site preparation costs. Access roads, weight limits, and frost laws can also be important in site development.

A sanitary landfill used for codisposal is normally owned and operated by an agency other than the wastewater treatment plant from which the

Table I. Design Criteria [1]

Method	Sludge Moisture Content (%)	Trench Width (ft)	Bulking Required	Bulking Agent	Bulking Ratio[a]	Cover Thickness Interim (ft)	Cover Thickness Final (ft)	Imported Soil Required	Sludge Application Rate (in actual fill areas)	Equipment
Narrow Trench	80-85	2-3	No[b]	—	—	—	2-3	No	1,200- 5,600 yd³/ac[c]	Backhoe with loader, excavator, trenching machine
	72-80	3-10	No[b]	—	—	—	3-4	No		
Wide Trench	72-80	10	No[b]	—	—	—	3-4	No	3,200-14,500 yd³/ac	Track loader, dragline, scraper, track dozer.
	72	10	No[d]	—	—	—	4-5	No		
Area Fill Mound	80	—	Yes[d]	Soil	0.5-1 soil: 1 sludge	3	3-5	Yes	3,000-14,000 yd³/ac	Track loader, backhoe with loader, track dozer
Area Fill Layer	85	—	Yes[d]	Soil	0.25-1 soil: 1 sludge	0.5-1	2-4	Yes	2,000- 9,000 yd³/ac	Track dozer, grader, track loader
Diked Containment	72-80	—	No[c,e]	Soil	0.25-0.5 soil: 1 sludge	1-2	3-4	Yes	4,800-15,000 yd³/ac	Dragline, track dozer, scraper.
	72	—	No[d,e]	Soil						
Sludge/Refuse Mixture	97	—	Yes[d]	Refuse	4-7 tons refuse: 1 wet ton sludge	0.5-1	2	No	500-4,200 yd³/ac	Dragline, track loader.
Sludge/Soil Mixture	80	—	Yes	Soil	1 soil: 1 sludge	0.5-1	2	No	1,600 yd³/ac	Tractor with disc, grader, track loader.

[a] Volume basis, unless otherwise noted.
[b] Land-based equipment.
[c] 1 ft = 0.305 m; 1 yd³ = 0.765 m³; and 1 acre = 0.405 ha.
[d] Sludge-based equipment.
[e] But sometimes used.

sludge originates. Under Michigan Act 64, P.A. of 1979, permits and other regulatory requirements are the responsibility of the landfill operator. This includes site selection and design, leachate control and operating techniques.

Codisposal Landfills

Codisposal is the mixing of sewage sludge with solid waste prior to disposal in a sanitary landfill. Many wet sludges are not suitable for sanitary landfilling; however, under normal conditions, a sludge dewatered to an 80% or less moisture content can be landfilled. The Michigan Department of Natural Resources (DNR) would prefer a moisture content of 60% or less, but this percentage is beyond the capability of nearly all processing facilities in the state. Figure 1 [2] shows the general operation of a sanitary landfill.

If the moisture content of a sludge is less than 70%, the material can be added directly into the landfill. Spreading with some refuse mixing is prefer-

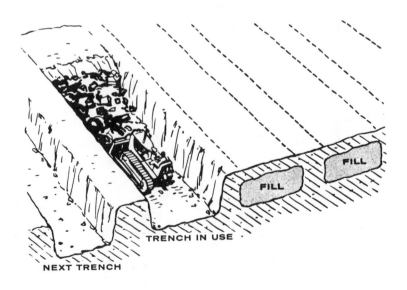

Figure 1. Sanitary landfill [2].

able to avoid large pockets of sludge in the fill. The sludge can be dumped near the operating face of the fill so the operator can mix portions of sludge into the stream of refuse. This ensures a reasonable sludge—refuse mix and proper sludge distribution in the fill.

If the moisture content of a sludge is greater than 70%, refuse mixing or sludge spreading is mandatory. Sludge and refuse can be mixed in a specially designated area or at the operating face of the fill. When mixing is done in a designated area, the sludge-refuse mix must be moved to the operating face for disposal. When the sludge is delivered directly to the operating face of the fill, the sludge is spread evenly over the layer of refuse and then mixed with additional refuse. The tracking and movement of the landfill equipment over the fill works the sludges into the refuse.

Several sanitary landfills have used mixed sludge and refuse as fill material. A landfill in Benton Township near Benton Harbor, Michigan landfills waste-water sludges from the joint treatment plant serving St. Joseph and Benton Harbor. Initially, the sludge was stored in piles on the site because the moisture content was too high for direct incorporation into the fill. The sludge was left to dry for a 30 to 60-day period and was then landfilled. The operation was conducted with difficulty, but no odor complaints were received. This landfill has no operating plan for several years, but it is now close to completion. As part of the closure plan filed with the Michigan DNR, waste treatment sludge is spread on the ground to dry and mixed with imported soil by bulldozer prior to application as final cover. This is being done to increase the humus content of the soil to enhance the cover crop.

Riverview, Michigan is constructing a high-rise sanitary landfill designed to extend 200 feet above the original grade. The fill receives more than 1000 ton/day of mixed domestic and industrial refuse. For a short period of time, wastewater treatment sludge was mixed with soil in an attempt to increase the organic content of the final cover. This was intended to aid the growth of a vegetative cover. The experiment was abandoned in under six weeks because mixing the sludge and soil into an adequate topsoil-like cover material significantly increased costs and slowed operations.

The Hagman Road landfill in Erie Township, Monroe County, Michigan accepted catsup and tomato waste from canneries in the Toledo, Ohio area. The moisture content was probably in excess of 98%. It was combined with refuse and soil before being incorporated into the landfill. The soil, refuse and tomato waste were mixed in a 30-foot square with 2-foot-high dikes. The wastes were mixed until the mixture had sufficient consistency for fill application. At this point, the dike on one side of the square was destroyed and the material bulldozed into the landfill. The tomato wastes were added to the fill after normal operating hours during the canning season in late summer and early autumn.

Sludge-Only Landfills

Sludge-only landfills can be constructed using any one of three basic techniques: trenching, area filling or diking. Trenching can be performed with narrow or wide trenches; area filling can be performed by mounding or layering. The selection of a particular construction technique is determined by site characteristics and sludge moisture content. Table II compares the effectiveness of various fill techniques under different site conditions and sludge moisture contents.

Both narrow and wide-trench sludge landfills are illustrated in Figure 2. The narrow trenches are less than 10 feet wide and are used for sludges with moisture contents of 85% or less. Final cover for the trench is usually 3-4 feet of earth taken from stockpiles left by the initial trenching. Narrow trench landfilling is particularly well suited for sites with steep slopes. A relatively large amount of land is unusable because the equipment must work from undisturbed earth.

Wide-trench landfilling uses trenches 10–40 feet wide. This technique is used for sludges with moisture contents below 80%, and preferably below 75%, because operating equipment must be moved over the filled sludge to spread, compact and cover it. When the moisture content of the sludge is too high, the tracks of drag lines are often equipped with pads. Final cover on wide trenches ranges from 3 to 5 feet, also taken from stockpiled earth.

The area fill technique is best used with sludges having a moisture content lower than 80%. The sludge can either be mounded in individual cells or alternately layered and covered (Figure 3). The layers or mounds are normally covered with 12–18 inches of soil. Normally, for both mounding and layering, the sludge is mixed with soil to enable vehicles to operate on it. Soil admixtures have varying makeups, depending on the type of soil, the moisture content of the sludge, and the workability of the admixture. Operating equipment must be furnished with wide tracks to facilitate movement on the relatively moist sludge. Area filling is best suited to areas with fairly level terrain. Final cover should be between 3 and 4 feet.

Dike disposal is a modification of the area fill technique. A dike is constructed around a level area, filled with sludge and covered with about 3 feet of soil. Moisture content in diked fills should be less than 80%. Soil addition ranges from no mixing to as much as four parts of soil to one part of sludge. Typical parameters for a diked disposal area are illustrated in Figure 4.

Albion, Michigan operates a secondary waste treatment plant on the Kalamazoo River. The sludge is digested and vacuum filtered. The filter cake is trucked to a sludge-only area landfill located on treatment plant property. No operational problems, leachate or odors have been reported.

Table II. Sludge and Site Conditions

Method	Sludge Moisture Content (%)	Sludge Characteristics	Hydrogeology	Ground Slope
Narrow Trench	72–85	Unstabilized or stabilized	Deep groundwater and bedrock	<20%
Wide Trench	>80	Unstabilized or stabilized	Deep groundwater and bedrock	<10%
Area Fill Mound	>80	Stabilized	Shallow groundwater or bedrock	Suitable for steep terrain as long as level area is prepared for mounding.
Area Fill Layer	>85	Unstabilized or stabilized	Shallow groundwater or bedrock	Suitable for medium slopes, but level ground is preferred.
Diked Containment	>80	Stabilized	Shallow groundwater or bedrock	Suitable for steep terrain as long as a level area is prepared inside dikes.
Sludge/Refuse Mixture	>97	Unstabilized or stabilized	Deep or shallow groundwater or bedrock	<30%
Sludge/Soil Mixture	>80	Stabilized	Deep or shallow groundwater or bedrock	< 5%

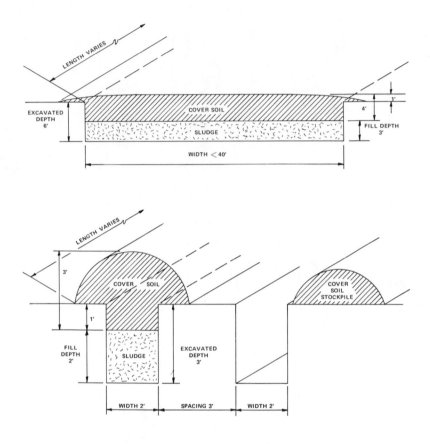

Figure 2. Typical trench landfill design [2].

Hastings, Michigan performed narrow-trench landfill of sludges for a period of time before going to landspreading. The experience there was that moisture content was too high in the lime polymer-centrifuged sludge. After a year in the ground, the moisture was still bound by the polymers and the land was not trafficable. Landfilling was abandoned after about a year.

LAND RECLAMATION

Reclaiming disturbed land through sludge disposal has been performed experimentally in a number of areas. Using sludge for land reclamation of strip mine tailings, deep mine tailings or abandoned gravel pits adds organic material to soils otherwise nearly sterile. Sludge can also improve the moisture-holding ability of a soil, sand or gravel base.

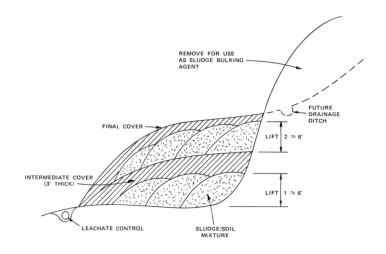

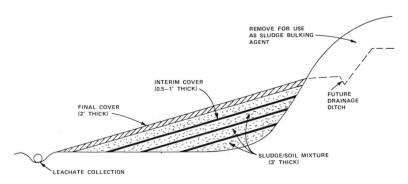

Figure 3. Typical area fill design [2].

An illustration of the need for improved organic content in disturbed land is that after 100 years vegetation is still absent on the gold mine tailings around Fairbanks, Alaska. Similarly, some coal mine tailings in southeastern Ohio are without vegetation after 50 years.

Sludge has not been used for reclamation in these areas because the mining industry is not equipped to spread sludge and has little economic incentive to use it to reclaim the land. The industry has permitted experiments but has not been able to justify large-scale programs using sludge to reclaim ravaged land.

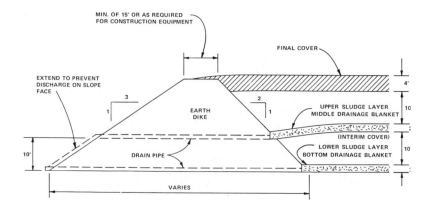

Figure 4. Typical diked landfill design [2].

SUMMARY

The problem of disposing of sludge will remain a number of years. In using landfills or reclamation for disposal, a community must carefully assess all aspects of the operation. Difficult terrain characteristics, hydrogeological and sludge cake characteristics all enter into an evaluation of sludge disposal.

Codisposal appears to be the preferred method of disposal. The advantages include little regulatory delay in approving the program because the site is usually in operation. Because the site already exists, there will be less impact from odors, traffic, aesthetics and other concerns than will likely be found in a new site using sludge only. Public opposition to the operation will often be less than for a sludge-only landfill. Using sludges in landfills can increase odor, leachate and operating problems. Nevertheless, landfilling sludge is a viable disposal option.

REFERENCES

1. U.S. Environmental Protection Agency, *Process Design Manual for Municipal Sludge Landfills*, EPA 625/1-79-011 Office of Technology Transfer, Washington, D.C. (1979).
2. U.S. Environmental Protection Agency, *Sludge Treatment and Disposal*, Vol. 2, EPA 625/4-78-012, Office of Technology Transfer, Washington, D.C. (1978).

CHAPTER 11

AESTHETIC RENOVATION

John W. Campbell, L. A.

Site Planning Development Incorporated
Charlevoix, Michigan

The Medusa Cement Company of Charlevoix, Michigan has been following an active Mine Reclamation Program since 1972. Until 1975 most of the work involved preliminary planning and research development of long-range plans, goals and objectives. These plans are intended to exceed the requirements of the Mine Reclamation Act (1970 P. A. 92). This foresight in management will make Medusa Cement Company a forerunner in mine reclamation in Michigan, and in the United States as a whole.

Prior to 1979 the Charlevoix Sewage Treatment Plant placed liquid sewage sludge in drying beds and trucked the dried material to an adjacent landfill. Through the cooperation of the city of Charlevoix, the Charlevoix Township Supervisor, District Health Department No. 3, Michigan Department of Natural Resources and the Medusa Cement Company, Site Planning Development, Inc. was authorized to implement a land reclamation program utilizing this sludge.

Sludge can be a valuable resource; however, it is not utilized actively as often as it is discarded as a waste product. Land application can be environmentally safe and beneficial to land reclamation and revegetation. However, when sludge is used for land reclamation, guidelines must be followed to establish application rates. A typical analysis of the sludge is presented in Table I. The program allowed 3000 gal/ac per application of liquid digested sludge and an annual limit of 18 dry ton/ac.

Table I. Sludge Analysis and Application Rate

Sludge type:	Wet sample, report prepared by Dr. Lee Jacobs, Michigan State University

Sludge sample results:	Average

Sludge Parameter	Concentration	Sludge (lb/dry ton)
Dry Solids, %	6.2%	----
Total Nitrogen	6.21%	----
NH_4-N	1.51%	----
NO_3-N	0.04%	----
Total Phosphorus	3.62%	72
Total Potassium	0.18%	3.6
Lead	640 ppm	1.3
Zinc	2400 ppm	4.8
Copper	1800 ppm	3.6
Nickel	90 ppm	0.18
Cadmium	54 ppm	0.11
Chromium	640 ppm	1.3

Available Nutrients/Ton of Dry Sludge

Nitrogen: available for plants	1st year	50	lb/dry ton
	2nd year	2.3	lb/dry ton
	3rd year	2.2	lb/dry ton
	4th year	2.1	lb/dry ton
Phosphate		170	lb P_2O_5/dry ton
Potash		4	lb K_2O/dry ton

Annual Maximum Sludge Rate Based on Cadmium

18 dry ton/ac

THE RECLAMATION PROJECT

Site Location and Soil Conditions

The Medusa Cement Company is located in Charlevoix County adjacent to Lake Michigan and the city of Charlevoix (Figures 1-3). The material to be reclaimed is overburden removed to expose limestone and shale material. At some locations in the quarry the overburden is 40-45 feet deep. the material is trucked and used to cover precipitator dust on sites adjacent to

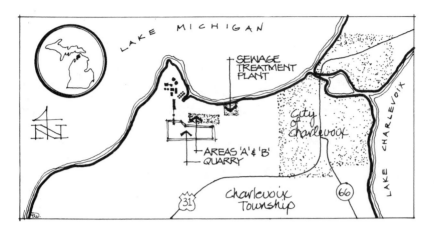

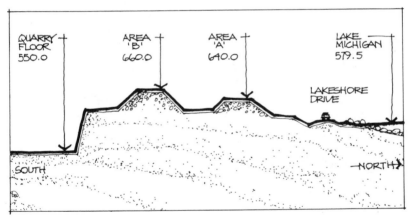

Figure 1. Site location and typical cross-sectional topography.

the quarry. The precipitator dust, waste from the cement production process, is covered with at least four feet of overburden. The overburden soil type includes sand, clay, shale and limestone, with very little humus topsoil (2 inches maximum). A typical soil pH is 8.4. Soil analyses indicate that the overburden soil contains negligible amounts of plant-available nitrogen and phosphorus, and relatively high amounts of plant-available potassium. This makes sludge an ideal amendment for these soils: sludge is high in nitrogen and phosphorus and low in potassium. Grading is designed to eliminate surface runoff toward Lake Michigan and toward the quarry floor.

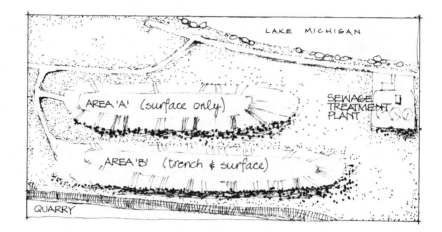

Figure 2. Map of Charlevoix site.

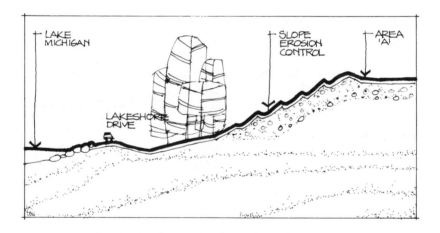

Figure 3. Sludge application location.

At the beginning of reclamation the overburden material has a high bulk density. A 42-inch subsoiler is used to loosen the soil (Figure 4). Farm equipment is useless as larger stone (12-60 inches in diameter) clutters the entire site at and below the surface. Although each subsoiler trench is 9 feet on center, it fractures the soil enough to allow moisture penetration, ridging and, eventually, the introduction of vegetation.

Grasses, legumes, wild flowers and trees are planted one year after preparation. The freeze–thaw action loosens the soil, and better results have been observed as a result of the waiting period.

TRENCHING

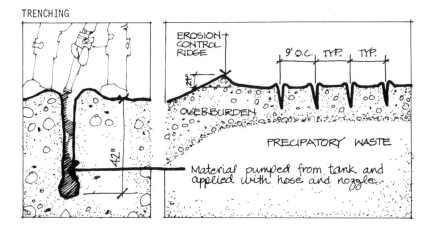

SURFACE

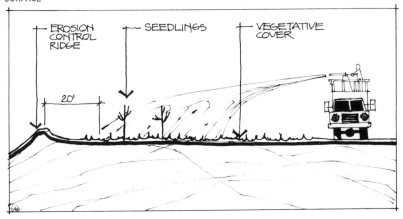

Figure 4. Subsoil and surface application methods.

SLUDGE APPLICATION

Sludge application procedures have been less of a problem than originally anticipated. Placing the sludge into the soil requires two operators: one to operate the tank truck and one to place the material, with a 2-inch hose and valve, in the slip trench. Surface application requires one individual to discharge the material.

The actual trench unloading time averages 20 min for 1500 gal of sludge. Surface application requires 12 min for discharge of 1500 gal and covers 0.5 ac. Approximately 20 min is required to load the vehicle. Hauling time is less than 5 min from the sewage plant to the site, as shown in Figure 2. Since it is presently impossible to incorporate the sludge into the soil after surface application, we adopted a surface application procedure. The hauling and application is accomplished with a Finn Hydroseeder and GMC 6 x 6 truck. This vehicle negotiates the rough terrain safely and efficiently.

In the spring, as soon as the site is accessible to heavy equipment, a D-9 bulldozer is used for subsoiling. Sufficient subsoiling trench is prepared to place sludge for one day's application, approximately 200 lineal feet. Under some soil conditions, one is limited to 1000 lineal feet of trench due to collapse later in the day. It required 17 days to complete the 8-acre sludge application. Immediately following the completion of this procedure this area was hydroseeded with grasses, legumes and wildflowers at rates shown in Table II. Immediately following seeding, the site was mulched at a rate of 0.5 ton/ac. Surface sludge application then began following germination of the grasses in July to encourage complete vegetative cover of the reclaimed area.

The average soil pH is 8.4; consequently, no lime is required. Our biggest problem is the lack of natural precipitation during the spring. Therefore, an irrigation system is used for six weeks to facilitate germination and establish vegetation. Prior to the surface application of sludge, the irrigation equipment is removed from the revegetated area. This keeps the equipment clean and prevents any accidental damage to the irrigation pipe.

TREE PLANTING

The present data are not sufficient to determine the long-range value of sludge and the effect on tree growth. One test area, completed in the fall of 1978 and observed during the 1979 growing season, showed favorable results. The plants used were Hybrid Poplar 206, American arborvitae (Michigan White Cedar), Norway Maple, Paperbark Birch, River Birch, Monarch Birch, European Mountain Ash, American Ash and Red Oak. All plants were one-year seedlings, which had been hand planted, as shown in Figure 5. Observation showed no negative effects and growth, as shown on Table III.

The use of cereal rye and its subsequent growth provided protection for the tree seedlings during the growing season from wind and soil temperature variations. Of 50 plants planted one Red Oak and one Monarch Birch died, for a 96% survival rate. By the end of 1984, it will be possible to make a final determination of the positive effect of sludge on the tree planting

Table II. Seeding Rates and Dates

Area A	Date of Hydroseeding: September 1978	
	Description	lb/ac
	Cereal Rye	40
	Pennlawn Fescue	20
	Baron Bluegrass	20
	Nuggett Bluegrass	20
	Manhattan Perennial Rye	20
	Conwed® Mulch[a]	1100
	Cover Established October 15, 1978	
	Date of Hydroseeding: May 1979	
	Dutch White Clover	5
	Penngift Crownvetch	5
Area B	Date of Hydroseeding: June 1979	
	Description	lb/ac
	Pennlawn Fescur	20
	Baron Bluegrass	20
	Nuggett Bluegrass	20
	Manhattan Perennial Rye	20
	Wildflower Mix	2
	White Dutch Clover	5
	Penngift Crownvetch	5
	Verdyol Standard Mulch®[b]	900
	Cover Established July 15, 1979	

[a] Conwed is a cellulose fiber mulch. This was applied following seeding.
[b] Verdyol Standard Mulch is a combination straw, cotton and paper mulch. This was applied immediately following seeding.

program. After this period, the natural process of nutrient recycling should be established for sustaining vegetation.

VEGETATION ESTABLISHMENT AND GROWTH

All sludge treated areas (12 acres) had a complete vegetative cover by August of 1979. Critics of the sludge program were amazed at the meadow-like appearance of the reclaimed area and the lack of offensive odors. In fact, the positive response of these former critics is having a sustaining effect

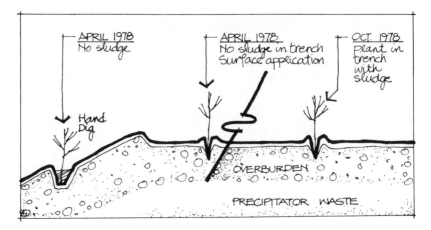

Figure 5. Planting of trees on overburden areas.

Table III. Tree Planting and Date

Date of Planting: April 1979 and October 1979

List of Trees Planted at a Rate of 450/ac

| Description | Growth in Height | |
	Without Sludge (in.)	With Sludge (in.)
Norway Maple	1- 5	1- 8
River Birch	1- 3	1- 4
Monarch Birch	1- 4	1- 5
Paperbark Birch	1- 3	1- 4
Poplar 206	1-18	2-24
American Arbovitae	1- 3	1- 4
American Ash	1- 6	1- 7
Red Oak	1- 3	1- 4

Plant survival, November 1979, greater than 90% for 5400 seedlings

on the program. Now it is time to wait and observe; however, from the initial program one sees only favorable results for the city of Charlevoix, the Medusa Cement Company and the recreation area.

GROUNDWATER AND SURFACE WATER

Due to the surface grading and land contouring, no change in water quality has been observed. As the subsurface is so compact, it is believed

Table IV. Sludge Cost per Acre and Commercial Fertilizer Equivalents[a]

The cost for the program is based on the following information:

Size of program:	12 acres
Total sludge volume:	45,000 ft^3 (337,500 gal)
Total dry material:	83.7 tons, or 7 ton/ac
(60 lb/ft^3 $\times$ 45,000 ft^3 $\times$ 6.2% dry material =)	

Nutrients and Their Dollar Value:

Nutrient	lb	Commercial Value[b] ($ 1979)
Nitrogen	4,740	3,550.00
Phosphate	14,230	5,340.00
Potassium	330	200.00
Total		9,090.00

[a] Actual cost of transportation and application of sludge both surface and subsurface, $13,863.00; Cost/ac, $1,155.00. Actual handling costs for the handling of material sludge, drying and hauling to landfill are not available. Approximate cost for equipment and manpower, $4,840.00. Cost based on 20 man-days, 5 days loader and 5 days truck.
[b] Value based on actual nutrients at retail prices.

that there will be very little moisture penetration below 4 feet from the surface.

VALUE OF THE PROGRAM

The cost of the sludge program is shown in Table IV. Money saved by Charlevoix and chemical equivalents received by Medusa indicate how successful this program is.

The city of Charlevoix has most of the sludge removed from the sewage treatment plant at no cost. The Medusa Cement Company receives the soil conditioning and increased vegetative response at a minimal increased cost over commercial fertilizer. Developing new cooperation between government and industry and producing an improved environment can only have a beneficial effect.

Not all reclamation sites are as conveniently located as in this program. The hauling distance is minimal; the equipment is locally owned and operated; the time schedules are flexible; the hazards are reduced; welfare of the community is improved; and the cost of the program is beneficial to all participating units.

CONCLUSIONS

Based on this program in Charlevoix, the following conclusions can be drawn:

1. Participation by all governmental units and corporate units can initiate a unique sludge disposal program.
2. The positive effects of the program must be analyzed and monitored yearly and recorded to ensure its continued success.
3. As the cost of commercial fertilizer rises, the proper use of sludge will be even more important to the reclamation industry.
4. As new equipment and application techniques are developed, the cost of application should decrease.
5. Efficient use of sludge can result in a successful revegetation program.

CHAPTER 12

LOW-RATE CROPLAND APPLICATION:
A CASE HISTORY IN JACKSON, MICHIGAN

Daniel C. O'Neill and W. William Farmer

Jackson Wastewater Treatment Plant
City of Jackson
Jackson, Michigan

Low-rate cropland application is the practice of applying stabilized sewage sludge to cropland at controlled rates. Applying sewage sludge to cropland is not a new concept: it has been practiced ever since the first production of sludge from treating sewage. What is different about the practice today, however, is the care and attention given to controlling the application rates. In addition, within the last few years specialized equipment has been developed and is available from several different vendors to apply sludge to the cropland. The improved methods of sludge management practiced today reflect both concerns expressed by the public and knowledge gained by scientists during the past decade in regard to safeguarding our environment. As a result, it is now possible to develop programs for applying stabilized sewage sludge to cropland with standards that basically will ensure that the nitrogen and phosphorus additions from the sludge will not exceed the amounts required by the growing crop, and that heavy metals present in the sludge will not accumulate to toxic levels in either the soil or the crop.

The prime legislative mandate for the establishment of environmental standards to govern the management of municipal sewage sludge is Section 405 of the Clean Water Act (PL 95-217). On September 13, 1979, the U.S. Environmental Protection Agency (EPA) published interim final rules and regulations for the application of sludge to land used for the production of food-chain crops [1]. Federal financial assistance, allocated in the Clean Water Act and administered by EPA, is available to qualified municipalities operating publicly owned treatment works (POTW) for planning, design and construction of improved sludge management programs that will comply with the regulations and protect our environment. In addition, National Pollutant Discharge Elimination System (NPDES) permits, under which municipalities operate their treatment works, require Michigan municipalities to submit to the state regulatory agency overseeing the operation of the treatment works an effective plan for effective sludge management. As a result of the Clean Water Act and NPDES permit program, all publicly owned treatment works are now required to implement an environmentally acceptable sludge management program.

SLUDGE MANAGEMENT PROGRAM

This chapter describes how the city of Jackson, Michigan developed its sludge management program, which, in part, includes low-rate cropland application of sludge. The Jackson Wastewater Treatment Plant is located in southeastern Michigan and serves a population of 90,000. The average daily flow of wastewater to the treatment works is currently 14 million gpd. Approximately 40% of the wastewater flow is contributed by industries, most of which are involved with the production or fabrication of metal parts for the automotive industry. The amount of digested sludge currently produced from treating the wastewater is 10 dry ton day. It is anticipated that the 1995 wastewater flow will be 19 million gpd.

After evaluating various sludge management plans on the basis of the factors listed in Table I, the city adopted a combination program involving the application of liquid digested sludge to cropland and the disposition of air-dried digested sludge in a sanitary landfill. Because of uncertainties about

Table I. General Factors To Consider In Developing A Sludge
Management Program

Energy Considerations	Reuse Potential
Labor Needs	Economic Feasibility
Chemical Requirements	Program Duration
Land Requirements	Environmental Acceptance
Operations Reliability	Public Acceptance
Operations Flexibility	

the future of energy costs and environmental regulations, the adopted program was designed on a ten-year basis. Approximately 40%, or 40,000 gpd of liquid sludge is applied to cropland during spring and fall. The combination-type sludge management program is quite cost-effective, very reliable and allows the reuse of a material too often considered a waste. It is in the interest of the reuse portion of the program—low-rate cropland application—that more detailed information is presented.

Before it can be determined whether a cropland application program is both feasible and acceptable, several specific factors must be considered in some detail. Two of the most critical are the availability of appropriate application sites and the acknowledgment of public acceptance. They are equally important and, for the most part, must be developed concurrently. That is, one cannot have a meaningful discussion on the concepts of cropland application of sludge with the public if there is no knowledge of potential application site; Conversely, one cannot investigate potential application sites without involving the public. With that in mind, a discussion on each of those factors, along with other practical matters affecting the overall development of the Jackson cropland application program, are presented.

PUBLIC PARTICIPATION

To have a successful cropland application program, it is vitally important to invite and receive public participation. In Jackson, this began with an initial survey of interest among area farmers. A candid explanation to the farmers on the intended cropland application program described the general characteristics of the sludge, pointing out its benefits and potential hazards. The type of equipment that would be used to apply the sludge was discussed, along with soil and crop monitoring. Finally, a breakdown of farmers' and city's responsibilities in implementing the program was explained.

Since the initial survey revealed an interest among area farmers in partici-cipating in a sludge reuse program, it was determined to hold a public meeting to explain the proposed cropland application program to all persons concerned. Table II lists several groups that were invited to the first public meeting.

Because sludge application sites were located outside the government's jurisdiction of the treatment works, it was important to discuss the program with the commissioners and supervisors representing the people living near the potential site locations and obtain their comments. Fortunately for the city of Jackson, it was possible to contain the sludge application sites within Jackson County. Obviously, for large-scale operations it may not be possible

Table II. Public Participation

City Commissioners	U. S. Soil Conservation Service (SCS)
County Commissioners	State Department of Natural Resources
Township Supervisors	Farmers
County Drain Commissioner	Cooperative Extension Service
County Road Commission	County Health Department

to limit application sites to one county, so public acceptance may be more difficult to achieve.

During the course of the sludge application program, followup meetings are held to update the public on the status of the program. The followup meetings provide an excellent opportunity to keep everyone in tune with the progress of the program and the concerns of its participants.

The goal of the public participation process is to ensure that all participating persons gain an understanding of the basic concepts of the cropland application program. If that goal can be achieved, any questions that might arise from the public can be responded to intelligently by the contacted agency or representative with prior knowledge about the program.

SITE SELECTION

A convenient and economical way of locating potential sludge application sites in Jackson was accomplished by an aerial survey of the farmland within a practical distance of the wastewater treatment plant. Potential sites spotted from the air were then cross-referenced with soil maps obtained from the district office of the SCS. The soil criteria used to determine whether a site qualified for sludge application are lisited in Table III, [2].

After the list of potential sites was reduced to a practical number, site visitations were made by the treatment plant staff. These provided much useful information in detailing final plans for cropland application. Once it had been determined that a farmer was interested in the sludge reuse program

Table III. Soil Limitations for Sludge Application to Cropland

Factor	Criterion
Slope	$< 6\%$
Depth to Water Table	> 4 ft.
Depth to Bedrock	> 4 ft.
Permeability	$0.6-2.0$ in./hr
pH	> 6.5
Available Water Capacity	> 0.1 in./in.

by the city and indeed had fields with appropriate soil characteristics, further details were explored. These discussions provided further information concerning the type of crops grown, existing soil fertility levels and irrigation capabilities in the farming operation. In most cases, corn is the crop grown on sludge-amended soils. After harvesting the corn, the grain is fed to beef cattle. The data available from the farmer's soil tests are used along with data from the city's monitoring program as input data in calculating sludge application rates for each site. The farmer's soil test data usually include the cation exchange capacity of the soil, soil pH, nutrient availability and possibly micronutrient (metal) availability.

The size of the fields is important for the Jackson sludge reuse program because of the extensive monitoring that is part of the program. Soil sampling and analysis are both time consuming and expensive. Therefore, the number of samples must be minimized or the cost of the operation becomes prohibitive. Consequently, a few large-acreage sites were chosen to monitor and control, rather than numerous small-acreage sites. The Jackson cropland application program currently involves two sites comprising a total of 400 acres.

From the site visitations, the exact distances from the treatment plant to the sites were determined. The application sites currently being utilized are approximately 12 miles from the Jackson treatment works. Also, the type of road classes on which the transport vehicles travel were determined from the site visitations. The type of road class is surprisingly important because sludge application in the spring of the year occurs while the Michigan frost laws are still in effect. The frost laws restrict the payloads of the transport vehicles. This particular problem has been averted by carefully selecting spring application sites to ensure access. In addition, each application site must have a convenient transfer location for pumping the sludge from the transport tanker to the application vehicle. Of course, this site must be off the highway and situated so that a public nuisance is not created.

Finally, the site visitations revealed the type of land use surrounding the application site, that is, the location of nearby wells and the type of housing and development that are prevalent in the area. Both sites currently being utilized in Jackson are quite rural.

After it was determined that a given site had a strong potential of becoming a chosen site, treatment plant staff conferred with the township supervisor and the county commissioner representing the government unit within which the site was located. Such contacts indicate the overlap between the site selection and public participation processes.

Having chosen application sites and gained public acceptance for the program, the city submitted the site selection information to the Michigan Department of Natural Resources for review and approval. Fortunately, all went well, and the city was permitted to implement its program.

APPLICATION RATES

Before sludge is applied to any of the approved sites, application rates are calculated. The purpose of establishing proper sludge application rates is to ensure the nondegradation of both soil and crops. In calculating application rates, it is first necessary to accurately determine the characteristics of the sludge to be applied, the characteristics of soils to which the sludge is to be applied, the nutrient requirements of the crop to be grown on the site, and the history of any previous sludge applications.

The sludge analysis should reveal: (1) the percentage of dry solids of the sludge; (2) the percentage of nitrogen, phosphorus and potassium in the sludge; and (3) the concentrations of heavy metals in the sludge, particularly cadmium, zinc, copper and nickel. Table IV presents typical data on the concentration of nutrients and metals in Jackson sludge.

Table IV. Current Jackson Digested Sludge Analysis

Constituent	Concentration
Total Solids	6.0%
Total Kjeldahl Nitrogen	5.0%
NH_4-N + NO_3-N	1.1%
Total Phosphorus	3.5%
K	0.5%
Ca	4.3%
Fe	7.7%
Zn	7250 mg/kg
Cu	1250 mg/kg
Cd	165 mg/kg
Cr	7500 mg/kg
Pb	200 mg/kg
Mg	7360 mg/kg

The soil analysis should reveal the available nitrogen, phosphorus and potassium, as well as the cation exchange capacity and the soil pH. Nutrient requirements for various crops can be obtained from the literature [3].

With the background information, the nitrogen available from sludge application can be computed. Because the sludge is injected into the soil, the available nitrogen from the sludge is equal to the total amount of inorganic nitrogen plus 20% of the organic nitrogen present in the sludge [4]. The expression for available nitrogen can be stated mathematically as in Equation 1.

$$\frac{\text{Available N lb}}{\text{ton sludge}} = \frac{(\% \text{ inorganic N}) + 0.20 \, (\% \text{ organic N})}{100} \times \frac{2000 \text{ lb}}{\text{ton}} \tag{1}$$

The available nitrogen from previous sludge applications must also be considered. A precise determination of available nitrogen from previous sludge applications cannot be accurately made because of unpredictable soil conditions that affect the rate of mineralization. Approximately 3% of the nitrogen from each year's previous sludge application is considered available in calculating application rates for Jackson sites.

An application rate based on nitrogen requirements can then be expressed as the crop requirement minus available nitrogen from previous applications divided by the available nitrogen in the sludge. The application rate based on nitrogen can be stated mathematically as follows:

$$\text{Application rate, } \frac{\text{dry ton}}{\text{ac}} = \tag{2}$$

$$\frac{\text{crop N req'd, lb/ac - available N from previous application lb/ac}}{\text{Available N in sludge, lb/ton}}$$

A similar procedure can be used for calculating an application rate based on phosphorus.

An application rate based on the amount of cadmium in the sludge must also be determined. For the first two years of the Jackson sludge application program, the annual cadmium limit was set at 2 lb/ac. That amount has now been reduced to an annual application rate of 1 lb/ac. This annual application rate is calculated in Equation 3:

$$\text{Application rate, } \frac{\text{dry ton}}{\text{ac}} = \frac{1 \text{ lb Cd/ac}}{\text{ppm Cd}/10^6 \times 2000 \text{ lb/ton}} \tag{3}$$

After calculating the various application rates based on the different influencing factors, the most limiting application rate is selected for sludge application.

In addition to the annual application rate for cadmium, a cumulative amount of cadmium is also established to protect a site's longevity. The present cumulative amount of cadmium allowed per site is 5lb/ac. Other micronutrients, such as boron, may also have to be evaluated to ensure no degradation of the soils or crops on the sludge application sites.

MONITORING

Even though sludge application rates are prudently calculated to prevent either soil or crop degradation, without a well-orchestrated monitoring program there can be little assurance that all will go as planned.

In view of the relatively high concentrations of heavy metals present in Jackson sewage sludge, it was particularly important to develop a monitoring program that would control sludge quality and prevent either soil or crop degradation.

Sludge Quality Control

For the city of Jackson, sludge quality control begins with the city's industrial waste pretreatment program. This work is actually performed by persons other than those directly involved in the sludge application program. However, through the efforts of those persons engaged in the prevention of excessive discharges of toxic substances, the quality of the sludge being applied to cropland receives not only its first level of protection, but also its most critical level of protection in terms of ensuring a long-term usable product. Table V shows anticipated sludge quality after anticipated industrial pretreatment.

Even with an active industrial waste pretreatment program, there is no guarantee that on occasion excessive amounts of toxic substances will not enter the treatment works, be incorporated in the sludge treatment processes, and contaminate a portion of the sludge to the degree that it would become unsuitable for cropland application. To prevent such an occurrence from going unnoticed by the treatment plant staff, sludge samples are collected daily from various points within the sludge-processing operations, composted and then analyzed for heavy metals on a weekly basis. The results of the analysis will reveal any adverse trends in the sludge quality. Problems can be

**Table V. Projected Heavy Metal Concentrations In Jackson Digested Sludge
After Industrial Pretreatment**

Element	Concentration (mg/kg)
Zn	800
Cu	300
Ni	300
Cd	50–100
Pb	300
Cr	1500

detected at an early stage, thereby preventing the application of contaminated sludge to cropland. In the event that a portion of the sludge is found to be contaminated, a management plan other than cropland application is implemented. At the Jackson Wastewater Treatment Plant, any digested sludge unsuitable for low-rate cropland application is diverted to large drying beds, where it is dewatered and ultimately disposed of in a landfill.

While performing sludge analysis from a given wastewater treatment plant, it is vitally important to ensure that the sludge analyses are accurate and meaningful. Because of the myriad of constituents comprising the sludge matrix, normal analytical procedures applied to wastewater analysis may lead to erroneous results when applied to sludge analysis. For this reason, the laboratory staff at Jackson Wastewater Treatment Plant took added measures to ensure the certainty of the sludge analysis data. Table VI shows the statistical data developed by the Jackson laboratory staff to ensure an acceptable level of confidence with the sludge analysis data.

Heavy metals are not the only constituents in sewage sludges that must be considered as potentially harmful substances from the practice of cropland application. Pathogens and organic chemicals must also be considered potentially harmful and included in the overall monitoring program.

Bacteria, viruses and parasites associated with sewage sludge will be affected to varying degrees by the anaerobic digestion processes and by injection into the soil environment [5]. Bacteria are the most fragile patho-

Table VI. Concentration and Percent Standard Deviation of Metals in Six Replicate Samples and Six 1.00 μg/ml Standards Digested with HNO_3 - $HClO_4$ -HF

Element	Samples			Standards		
	μg/g		% SD	μg/g		% SD
Fe	54,040	1,340	2.5	.89	.0098	1.1
Ni	1,348	39.4	2.9	.99	.0179	1.8
Cu	1,108	25.2	2.3	.97	.0160	1.6
Cd	445	5.4	1.2	.97	.0103	1.1
Zn	7,703	77.8	1.0	.96	.0052	0.5
Cr	9,033	238.7	2.6	.995	.0122	1.2
Pb	512	25.2	4.9	1.00	.0000	0.0

gens and are greatly reduced in numbers by anaerobic digestion. After injection into the soil environment, further competition reduces the number of pathogenic bacteria. Viruses are more persistent than bacteria, but exposure to sunlight and drying conditions will eventually inactivate them.

Parasites are the most resistant type of pathogens. Fortunately, the intermediate hosts necessary for most parasites to develop to an infectious stage are not readily available in the soil environment. Perhaps the most important parasite to be considered with sludge application is *Ascaris lumbricoides* because of its extreme resistance to adverse conditions and its worldwide prevalence [6].

Even though the possibility of disease transmission does exist from applying even digested sludge to cropland, there is no epidemiological evidence to date indicating that human illness has occurred from properly managed operations [5].

With regard to organic contaminants, there are few, if any, industrial sources of organic chemicals in Jackson. Still, Jackson sewage sludge has been analyzed by the Department of Natural Resources laboratory for selected toxic organics. The results were negative.

Field Monitoring

To prevent either soil or crop degradation, a field monitoring program complements the sludge quality control program instituted at the treatment works. Such a program ensures that heavy metals do not build up in the soil to damaging levels and that toxicants are not allowed to enter the food chain or damage the crops. In addition to the soils and crops, water filtering through the soil is sampled and analyzed in irrigated fields.

Soil monitoring begins before the first application of sludge. Soil samples are collected at various increments to a depth of 5 feet and analyzed for pH, available phosphorus, cation exchange capacity, available metals and total metals. This sampling provides the baseline data to refer to should any question arise regarding the impacts of sludge application.

Once sludge application is initiated on a field, annual soil testing is conducted, but to a lesser intensity than the background soil testing. The depth of sampling is reduced to 12 inches; however, the same type of analyses are performed. The data generated from the annual soil testing will be the first indicator of any changes in the soil characteristics. To facilitate interpretation of these data, a control area having received no sludge must be included in the sampling and analyses. Further, one must know what other materials, such as commercial fertilizer, have been applied to the site.

A portion of one of the application sites has been divided into narrow bands of soil that have received sludge once, twice, three times, etc. Samples from these areas are collected and analyzed each year to evaluate the effects of multiple sludge application and also to project future soil conditions once sludge applications are discontinued on a given site.

Table VII lists the concentrations of various elements that have been found in the control soils and sludge-amended soils after one year of sludge application. Sludge was applied at a rate equivalent to 2 lb/ac/yr of cadmium. After one application of sludge, no apparent differences could be seen in the concentrations of the elements between the control and sludged soils. Concentration differences of the elements in the sample must be greater than the differences caused by sample variability and analytical variances to be recognized as real in the data.

Plant monitoring consists of sampling different parts of the plant at various times during the growing season. While preparing the plant samples for analysis, one must be very careful not to contaminate the plant samples with the preparation equipment and be certain that the digestion procedure will produce accurate results.

The crop analysis should also include a control sample grown on non-sludge-amended soil, as well as the crops grown on the sludge-amended soils. The plant analyses include total metals (cadmium, zinc, nickel, copper, lead and chromium) and total phosphorus. Greater sensitivity will be required in the instrumentation used for analyzing for heavy metals in the plant samples because of the lower concentrations present in the prepared sample. When comparing data from plant analyses, it is important to record to which hybrid

Table VII. Concentration of Metals in Sludged and Control Soils

Sample	Depth (cm)	Concentration (mg/kg)				
		Cd	Cr	Pb	Cu	Zn
Control	0-5	0.31 (0.04)	32 (7)	5.2 (3.6)	22 (8)	63 (14)
	5-10	0.49 (0.24)	36 (9)	4.3 (2.4)	23 (9)	65 (10)
	10-15	0.33 (0.17)	34 (11)	4.8 (2.0)	21 (5)	57 (8)
	15-30	0.38 (0.09)	37 (5)	4.1 (0.8)	14	85 (27)
Sludged	0-5	0.45 (0.28)	53 (8)	4.7 (0.90)	23 (8)	49 (11)
	5-10	0.45 (0.25)	60 (6)	4.1 (0.70)	23 (3)	54 (9)
	10-15	0.58 (0.31)	66 (21)	5.6 (2.50)	21 (8)	60 (17)
	15-30	0.37 (0.21)	49 (8)	4.4 (0.40)	23 (6)	49 (6)

[a] ( ) denotes standard deviation.

the data pertain because there can be significant deviations in the amount of heavy metals taken up by different hybrids. Final crop samples should be collected and analyzed before the crop is harvested.

Table VIII lists the concentrations of various elements in corn grain samples taken from control and sludged areas. No significant differences in the concentrations were detected between the control and sludge-amended samples. Also, all concentrations fall within the ranges reported in literature for corn grain.

If a field is irrigated, water samples are collected after each irrigation cycle. Irrigation fields have lysimeters installed at depths of two feet and four feet located at various points in the field. Along with the water samples collected from the lysimeters, a sample of the irrigation water supply is also included in the analysis. The analyses performed on the water samples are: total metals; inorganic nitrogen, phosphorus, pH and conductivity. Table IX lists the concentrations of various elements found in the water filtering

Table VIII. Corn Grain Data

	Concentration				
Treatment	Cd (μg/kg)	Pb (μg/kg)	Ni (mg/kg)	Zn (mg/kg)	Cu (mg/kg)
Sludged	73	132	3.50	17.9	1.9
Standard Deviation	41	95	0.38	3.1	0.16
Range	21-140	53-317	2.5-3.8	15.0-25.5	1.8-2.3
Literature [7]	35-1200	30-500	0.1-5.0	12-100	0.9-17.0
Control	57	138	3.3	17.8	2.4
SD	40	59	0.8	3.8	0.8
Range	44-103	75-216	2.0-4.8	12.8-23.5	1.8-3.8

[a] Trojan hybrid

Table IX. Lysimeter Data (1979)

	Concentration					
Treatment	Cd (mg/l)	Ni (mg/l)	pH	Cond. (μmhos)	PO_4-P (mg/l)	NH_4-N +NO_3-N (mg/l)
Sludged, 2 ft	0.012	0.052	6.3	954	0.059	76.9
Control, 2 ft	0.019	0.064	6.5	1500	0.191	74.0
Sludged, 4 ft	0.011	0.044	6.0	878	0.147	77.3
Control, 4 ft	0.010	0.038	6.5	611	0.259	40.5

through the soil. Note the high concentrations of inorganic nitrogen appearing in all of the samples, which is probably indicative of excessive commercial fertilization.

EQUIPMENT SELECTION

The care and thought exercised in determining sludge application rates and in setting up a thorough monitoring program must be supported by dependable equipment to make the cropland application program completely functional.

Equipment selection for the program essentially involved two types of equipment: field equipment, consisting of trucks and other vehicles used to transport and apply the sludge to the cropland; and laboratory equipment, consisting of apparatus and instrumentation needed to analyze the sludge, soil and plants.

The main pieces of field equipment purchased by the city were two 5500-gallon stainless steel transport trailers, three tandem trucks to pull the trailers, and one terra-tired sludge applicator with three injection tubes. In selecting the field equipment, particular attention was given to corrosion resistance. For this reason, stainless steel tanks were specified. Also, there are standby pieces of equipment for those items most likely to have operational failures. For example, an extra tandem truck is available to maximize application efficiency by minimizing equipment downtime. The tandems are powered by a 290 Cummins diesel coupled with a ten-speed Fuller transmission and 4.10 rear axle. The sludge applicator is equipped with the terra tires to minimize soil compaction, an important aspect of good farming practices. By injecting the sludge into the soil, potential odor problems are eliminated; the loss of ammonia nitrogen through volatilization is significantly reduced; and the soil is aerated by the chiseling effect of the tines as the injector knifes through the soil. An inventory of replacement parts for critical components, such as the applicator pump, is maintained. A van is also available to make site inspections and collect samples.

The major pieces of laboratory equipment currently being utilized for the monitoring work are listed in Table X.

PERSONNEL

The manpower requirements for performing the work of the Jackson sludge reuse program were less intensive than many other types of sludge management. The same personnel carrying out the cropland application pro-

Table X. Laboratory Equipment

Laboratory Item	Use
Atomic Absorption Spectrophotometer	Heavy metals analysis
Carbon Analyzer	Organic carbon analysis
Polarographic Analyzer with SMDE	Heavy metals analysis
Kjeldahl Distillation Apparatus	Nitrogen analysis
Spectrophotometer	Phosphorus analysis
Rotating Kjeldahl Digestion Apparatus	Sample preparation
Soil Grinder	Sample preparation
Wiley Mill	Plant sample preparation
Receprocating Shaker	Extractable elements

gram in the spring and fall are also responsible for landfilling air-dried sludge during the summer months. The responsibilities and duties of the sludge application program are shared by three levels of personnel employed at the treatment works. Part of the superintendent's time is spent directing the program. A full-time soils scientist is employed to supervise and monitor the operations and two full-time equipment operators are employed to apply the sludge to the cropland and maintain the equipment.

The responsibilities of the superintendent as they relate to the sludge reuse program include: (1) setting policy on the manner in which the program should be carried out; (2) establishing a positive public relations program; (3) reviewing the data generated from the monitoring program; (4) preparing an operating budget for the fiscal year; and (5) examining the long-term cost-effectiveness of the program. The Superintendent's work is reviewed and discussed with the city manager. Final budget adoption and long-term planning are ratified by the City Commission.

The soils scientist supervises the actual sludge application operation by calculating application rates for a given site and scheduling the time of sludge applications. As part of the operation, the soils scientist is also responsible for monitoring sludge quality within the treatment works and for establishing a field monitoring program, which includes the sampling and analysis of soils and crops from the various application sites. In addition, the soils scientist is responsible for setting up research studies on low-rate cropland application and interpreting all data generated from the monitoring programs.

The two equipment operators must be qualified truck and trailer operators who demonstrate good judgment and proficiency when handling the equipment. In addition to operating the sludge transport and application vehicles, the workers are required to maintain the equipment. This would include all necessary preventive maintenance and repair work required to keep the vehicles in good operating condition, with the exception of major engine and transmission overhauls. The equipment operators are also encouraged to become familiar with the technology used in low-rate cropland application so that they will be more knowledgeable of the overall concepts of the program. This knowledge will not only help them discuss the operation more intelligently with the farmers receiving the sludge, but will also make their job more interesting. At least one other treatment works employee is familiar with the operation of the sludge applicator so that a fill-in equipment operator is available.

COSTS

Differentiating the costs of the cropland application program from the total costs of the overall Jackson sludge management program is difficult at best, and really does not reflect the true costs for cropland application programs in general. The reason for this perplexity is hidden in the fact that equipment and personnel utilized in implementing the cropland application program are also utilized in landfilling air-dried sludge.

However, keeping the duality of functions in mind, an effort has been made to identify the various costs of the cropland application portions of the total program. Capital costs (i.e.) are essentially equipment and operation and maintenance (O&M) costs, are shown in Table XI. The unit cost per ton of dry sludge is based only on the O&M costs. All costs are projected 1980 costs.

AREAS OF FURTHER RESEARCH

An effective monitoring program not only should have the capacity to recognize immediate problems as they arise, but should also answer questions that may have a vital impact on future sludge disposal. Areas of concern either being looked at now or to be examined in the future are as follows:

1. What is the movement of metal in soils and accounting for the quantity of metal placed in the soil through close increment sampling;
2. What is the cadmium uptake by corn as affected by zinc concentrations in the plant;
3. What effect will high phosphorus concentrations in sludge have on plant growth;
4. What is the nitrogen availability from Jackson sludge and application sites.

Table XI. Costs of Cropland Application of Sludge

Capital Costs	
Tandem trucks	$100,00.00
Tank trailers	50,000.00
Applicator	50,000.00
Van	5,000.00
Instrumentation and apparatus	50,000.00
	$255,000.00
O&M Costs	
Personnel	37,000.00
Superientendent, 400 hours	
Soils scientist, 2,000 hours	
Equipment operators,	
2,160 sludge application	
640 maintenance	
Fuel, 16,000 miles @ 4 mpg	6,200.00
4,160 gal. @ $1.50/gal.	
Equipment maintenance	4,500.00
Supplies	3,000.00
Contingency	3,000.00
	$ 53,700.00
Unit Cost	
Application days, 5-month period	90
Trips, 5,300 gallons each	700
Sludge applied	850 Tons
Cost per dry ton,	
$53,700/850 ton	$ 63.00

Research on the first two items is now being conducted through greenhouse and field studies. Preliminary data on close increment sampling indicates that this method can be used to account for the amount of metal being applied if injection is the method of application. Soil samples were taken one inch apart in a sawtooth pattern across a plot seven feet by two feet perpendicular to the direction of sludge application. The analyses revealed a profile across the trenches made by the sludge application. The distance between the sludge injection tubes is 24 in. center-to-center. It is thought that this sampling technique yields a more accurate account of the concentrations of metals applied. Figures 1 and 2 show close-increment sampling data for zinc and nickel analyses. Note that the movement of nickel through the soil is more pronounced than that of zinc. The movement of the metals is depicted by the different slopes and amplitude of the peaks.

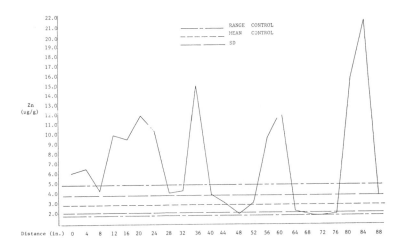

Figure 1. DPTA extractable zinc close-increment soil sampling.

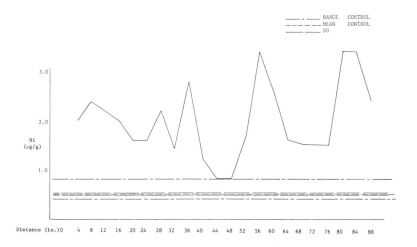

Figure 2. DPTA extractable nickel close-increment soil sampling.

SUMMARY

To have an environmentally acceptable sludge management program, the city of Jackson, Michigan has established the practice of low-rate cropland application as part of its overall sludge management program. This program has not only provided the city with an economical means of managing a significant amount of its wastewater sludge, but has also provided participating farmers with valuable crop nutrients and soil conditioners. The city performs extensive monitoring as part of the program management to protect against any soil or crop degradation.

Detailed discussions presented on the development and implementation any of the program show the importance of public participation, criteria for site selection, calculation of sludge application rates; development of a monitoring program; selection of equipment; responsibilities of the personnel; and costs of the operation.

Finally, areas of further research are presented in an effort to answer questions that may have a vital impact on future sludge disposal.

REFERENCES

1. "Criteria for Classification of Solid Waste Disposal Facilities and Practices," *Federal Register* 44, (179): 53460 (1979).
2. Wisconsin Department of Natural Resources. "Guidelines for the application of Wastewater Sludge to Agricultural Land in Wisconsin," Technical Bulletin No. 88, Madison, WI (1975), p. 24.
3. "Fertilizer Recommendations for Vegetables and Field Crops," Extension Bulletin E-550, Michigan State University, E. Lansing, MI (1976). p. 3.
4. Jacobs, L.W. "Sewage Sludges--Characteristics and Management," in *Utilizing Municipal Sewage Wastewater and Sludges on Land for Agricultural Production*, L. W. Jacobs, Ed. North Central Regional Extension Publication No. 52, Michigan State University, E. Lansing, MI (1977), p. 15.
5. Pahren, R. H., J. B. Lucas, J. A. Ryan and G. K. Dotson. "Health Risks Associated with Land Application of Municipal Sludge," *Water Pollution Control Fed.*, 51: 2588 (1979).
6. Fitzgerald, P. R., and R. F. Ashley, "Differential Survival of *Ascaris Ova* in Wastewater Sludge," *Water Pollution Control Fed.* 49: 1722 (1977).
7. Pietz, R. I. et al. "Variability in the Concentration of Twelve Elements in Corn Grain," *J. Environ. Qual.* 7:106-110 (1978).

CHAPTER 13

SEWAGE SLUDGE COMPOSTING
AT THE BLUE PLAINS
WASTEWATER TREATMENT PLANT,
WASHINGTON, DC

Francis A. Riddle

District of Columbia Department of Environmental Services
Washington, DC

The Blue Plains Wastewater Treatment Plant in Washington, DC treats the wastewater from that city and much of its suburban area. Between 20 and 30 years ago, approximately 100 ton/day of digested sludge at 20% solids were produced. It was left in an open field to dewater further and to age for about a year, then was given away to all comers for soil conditioning, lawn dressing, etc. Any possible impact on the health of sludge users and on the environment was not considered to be a problem. But the area grew and the plant had to be expanded and upgraded, some phases of which are still underway. Today, the plant is a 309 million gpd plant and produces approximately 1300 ton/day of sludge at 20% solids. Of this quantity, 350 tons are digested and the remainder is undigested or raw. But in a few years, as facilities still under construction are completed and go online, sludge quantities are projected to reach 220 ton/day! Needless to say, sludge disposal has become the number one problem. At present, the digested sludge is spread on agricultural, public or marginal lands. The application rate on agricultural lands is approximately 50 wet ton/ac and can go as high as approximately 250 ton/ac on marginal land. In either case, it is either disced or plowed under; at least, it is utilized.

Raw sludge is a different matter. The bulk of it is disposed of by trenching. This is an interim disposal method in which a trenching machine digs a trench 2 feet wide and 3 feet deep. The trench is filled with the sludge and immediately covered as the machine digs the next trench. This is very costly, not only in operating dollars, but in equipment and land usage as well. Further, very little, if any, benefit is derived from the sludge.

In the search for a better way, Blue Plains has undertaken a large-scale project to compost approximately 300 ton/day of raw sludge using the forced-aeration, static pile method developed by the U. S. Department of Agriculture (USDA) of Beltsville, Maryland.

STATIC PILE COMPOSTING

The Site

The composting site is a 14-acre area located on the treatment plant grounds. The nearest residence is 1800 feet away. Each of four aeration pads is approximately 1.2 acres in area; two curing pads and the mixing area are approximately 1 acre each; and the remainder is taken up with roads, leachate ponds, scrubber piles and blowers. Only two of the aerating pads and the mixing area are paved, and the whole operation is open to the weather. The site is operated by a minority firm under contract to the District.

Description of the Process

In the Beltsville method, sludge is mixed with a bulking agent. In this case, wood chips are used in a ratio of approximately 2:1 by volume. A mobile mixer is the primary mixing device, with front end loaders as backup. The resulting mixture is placed on a bed of wood chips into which 5-inch perforated plastic pipe has been placed. The pile is built to a height of approximately 10 feet and covered with a 1-foot-thick blanket of previously composted material. The 5-inch perforated pipe is connected to a length of solid pipe that extends out of the pile, which is connected to a 10-inch header. Two such lines, on 8-foot centers, are connected to each header, which is connected to a blower. The blower is a centrifugal type powered by a 2-hp motor. Normally, the blowers are controlled by a timer on 15 -minute on-off cycles. The discharge is directed into a scrubber pile of finished compost, which absorbs any odors. At the end of the aeration stage of at least 21 days, the moisture content is generally in the range of 55-58% moisture. A recent trial was run in which one 8-inch line was connected to each blower and the blower run almost continuously in an attempt to achieve a greater moisture reduction. However, no significant improvements were noted.

After aeration, the compost is transferred to a curing pile and left to cure for at least 30 days. After curing the material is screened.

Screening

At Blue Plains, screening of the finished compost is utilized primarily as a matter of economics to reclaim the chips for reuse. When composting got underway in February 1979, a contract in excess of $650,000 was awarded to purchase 83,000 cubic yards of wood chips. Needless to say, it is imperative to reclaim the chips and equally important that the screen itself do as little damage as possible to the chips. Screening is also important depending on the end use of the final product. If bulking agent is cheap and readily available and the compost is to be used on a grand scale in reclamation work, agriculture or nursery applications, screening may not be necessary.

The screening system at Washington consists of a loader, conveyor and the screen itself. The loader is a homemade unit that holds approximately 10 cubic yards and was originally intended as the interface between a dump truck and a sludge trench. However, it was never used for this purpose and was modified to process compost. The unit is equipped with a manure spreader type of unloading mechanism of chains and flights dragging across the bottom, which carry the compost to a chute where it drops onto a conveyor. The handling of compost when it is less than 47-48% moisture is no real problem; however, as the moisture content approaches 48-50%, handling does begin to get difficult. In this unit, the moist compost compacts and builds up under the chains and flights to a point where a pick and shovel are required to clear it out. At first, the unit was a fairly high maintenance item, but as the bugs were gradually worked out, the unit has performed satisfactorily.

Two different screens were tried, The first was a fairly small, self-contained mobile unit that was obtained on a rental basis. It was a double-deck unit mounted at the end of an inclined conveyor. The screen had been used successfully in many applications, including sand and gravel work and coal mining operations. It was a well-engineered unit, diesel powered and hydraulically driven. The actual screening unit consisted of a double deck with both a coarse- and fine-mesh screen. The deck panels are changeable and numerous combinations of screens were tried, including square, oblong and diamond-shaped holes, and harp wire. All worked with varying degrees of success if the material was under 48-50% moisture. However, the fine screen eventually clogged and the machine design was such that it was difficult to get access to clear it. Blinding resulted in downtime and, of course, a decrease in output. Under normal conditions, that unit could process approximately 50 yd^3/hr; however it was difficult to sustain continued operation and its use was discontinued.

Eventually, a second, larger screen was installed—a Liwell® screen. Liwell is a unique unit in which urethane deck panels alternately stretch and flex in a rippling motion, which mechanically fluidizes the bed of feed material. The screen is actually a box within a box. One edge of the deck panels is attached to the cross members of the inside box and the other to cross members of the outside box. An eccentric shaft causes the two boxes to oscillate in opposite directions, which results in the stretching and flexing of the panels.

This unit can handle material that has a slightly higher moisture content than could be handled with the first screen that was tried. For screening purposes, the drier the compost the better, until the point where excessive dust is a problem. The Blue Plains experience is that best results are obtained with moisture content in the 40-47% range. Performance begins to taper off, although effective separation still is achieved in 48-50% range. Screening can continue in the 50-52% moisture range, but the feedrate must be cut drastically. It is not practical to screen when the moisture content is above 53%.

The screen is equipped with 26 of the flexible panels. A range of hole sizes is available and each panel can be replaced individually. The unit was delivered with all panels having 0.5-inch square holes. Most of the screening action took place in the first half of the screen. With the 0.5-inch size hole the full length of the screen, the long narrow chips (about the size of the sharpened end of a regular wood pencil) had many opportunities to fall through. Half the panels were then replaced with 0.25-inch hole panels. This produced a more uniform end product by presenting fewer opportunities for the long narrow chips to pass through.

One of the other desirable features of this screen is that it is almost self-cleaning. If it does become clogged, the feed is shut off and the screen left running. The stretch-flex action will clear the screen in a matter of minutes. This is a major factor as far as ease of operation is concerned.

Improvements to the Site

The Blue Plains plant is a regional plant serving the District of Columbia and large parts of suburban Maryland and Virginia. Naturally, sludge disposal is a regional problem and has been a sore point among the politicians for years, finally ending up in the courts. The construction of the composting site was the result of a court order with a deadline for beginning operations within 6 months. As a result, we are now going back and making improvements. One area receiving major attention is mixing. The present method is labor-intensive and subject to the weather. It requires the mobile mixer, two dump trucks, two front end loaders and several operators. An automated wood chip receiving, storage and mixing facility is under construction. The receiving bin is a 65-cubic yard bin designed to receive chips from front end loaders or dump trucks. It is a live-bottom bin discharging to an inclined lift conveyor. The conveyor transports the chips up a 60° incline and deposits them on a flat distribution belt conveyor, which distributes the chips evenly in the storage bin. This storage bin is 100 feet long, 15 feet wide and 30 feet high. It will hold approximately 1700 cubic yards, or two days chip requirement. The chips are retrieved from the bottom of the bin by a unique screw feeder/reclaimer. This unit is a Wennberg Parascrew®, which travels back and forth on fixed parallel rails under the stored wood chips. As the screw travels longitudinally, it also rotates, feeding material transversely to a retrieval conveyor. The feedrate can be regulated by varying the rotational speed of the screw. The retrieval conveyor transfers the chips to a drum mixer, where the sludge is introduced and mixed with the chips. The drum is 6 feet in diameter by 14 feet long and is rotated by a frictional drive powered by a 50-horsepower motor. Output can be varied by changing the drum angle of inclination, speed of rotation, angle and spacing of mixing blades or any combination thereof. The unit is designed for a maximum rate of 45 ton/hr of wet sludge. The mixer discharges onto a conveyor which carries the mix to the composting area. It is anticipated that one person will be able to operate this facility.

Problem Areas

At present, the main problem with the composting operation is difficulty in screening. Being exposed to the weather has made it very difficult to get material dry enough to screen during the winter months.

Several attempts have been made to reduce moisture by blowing ambient air through piles of cured material prior to screening. Also, aerating windrows with the mobile mixer has been attempted, but nothing has produced significant results so far. It appears that spreading compost in a 6- to 12-inch layer to air-dry is effective, but requires space and is very weather-dependent. This method has been tried recently but it is too soon to determine its large-scale applicability.

Cost

The bottom line of all this effort is cost. The capital costs to construct the facility were approximately $5 million. As of November 7, 1979, the operating costs were $23.44/wet ton, or about $117/dry ton. The two major cost items are the contract to operate the site and the wood chips.

SUMMARY

Composting may not be the answer to the sludge disposal problems of every community. Certainly there are problems and shortcomings with the site and operations, but it is felt that a good start has been made. As the problems are worked out and effort continues to improve this operation, there is confidence that a prominent place for composting in the total scheme of sludge disposal in the United States can be demonstrated.

CHAPTER 14

CALCINING SLUDGE—A PARTIAL SOLUTION

James C. Scott, P. E.
Utilities Department
Ann Arbor, Michigan

Developing an appropriate discussion about Ann Arbor's calcining facilities presents a real challenge. First, it is billed under the heading "Land Disposal." As one of the major objectives in calcining is to minimize or eliminate the need for land disposal, inconsistency exists here. Second, in terms of the ultimate disposal of sludge, we cannot claim to meet this objective. Significant residual quantities of waste remain after calcining. This chapter explains how Ann Arbor fits into this whole scheme and how its programs have been developed to keep pace with its changing needs.

Ann Arbor's current Water Treatment Plant Sludge Disposal Program had its beginnings in a facilities report prepared in 1962, when engineers recommended continuing the then current program of sludge lagooning. However, the city selected the alternative solution of calcining. The estimated cost in either case—$400,000. In 1969, after construction had been completed at a cost of $1.2 million, it was reported [1] that stable production in the calciner had finally been achieved; that the need for waste disposal was being reduced to minor amounts of magnesium hydroxide and other waste materials; and that Ann Arbor had taken a "wise and significant" step in reducing the waste problems facing it. Time has proved the first two assertions to be incorrect.

It was not until 1979 that all sludges from the water treatment plant were processed through the calciner (except during periods of planned maintenance). Those accustomed to allowing for 5 years from conception to completion, should take note of this 17-year lead time. And what of the "minor" amounts of magnesium hydroxide and other waste materials? By 1972 these materials had filled the existing lagoons and were spilling over onto the adjacent Michigan State Highway right-of-way. At this point, someone could easily have wondered how "wise and signifcant" a step had been taken.

The purpose of this dismal picture is to emphasize the need for continual attention to advanced planning, facilities evaluation, technological advancements and the changing needs of the public. There is a brighter side to this picture! Since 1969 Ann Arbor has had an active program directed toward responsible management of its sludge disposal needs. The Utilities Department in-house engineering staff assumed responsiblity for this program, aided by outside consultants and contractors.

In the original journal article in 1969, problems of personnel, intermittent operation, sludge collection, freezing, scrubber efficiency, lime slaking and spare parts had already been identified. Since that time, operating experience and specific study have disclosed additional problems, some of which are as follows:

1. Inadequate thickener capacity was constructed for the calciner. Based on design expectations, only 25% of the needed thickener capacity was constructed. Greater efficiency has been achieved in practice however, so that available capacity satisfies approximately 50% of the calciner's needs.
2. Chemical scaling of the high-speed exhaust fan impeller produced equipment imbalance. This aggravated maintenance problems and caused structural damage to the building.
3. Intense heat has caused major failure in the exhaust duct immediately downstream of the reactor.
4. Internal maintenance is a major factor, requiring reactor cooling every 800–1000 hours of operation. The duration of each shutdown is 4–7 days.
5. As a result of adverse environment and application, the service life of much of the equipment is shorter than anticipated, and *far* shorter than the service life normally encountered with water treatment plant equipment.
6. Sludge in the thickener must be "conditioned" through the scrubber (recarbonation and heating) before efficient calciner operation can be achieved. With only one thickener, periodic unwatering, inspection and repair have been repeatedly deferred because of this conditioning requirement.
7. Improper installation of original equipment caused inefficient operation of the thickener. This was not discovered until maintenance was being performed recently.
8. Inadequate building ventilation and dust control problems were encountered related to both the calciner and purchased chemicals.
9. Settleability of sludge in the lagoons was decreased by the high concentration of magnesium hydroxide. This aggravated the problem of already limited lagoon capacity.

10. The use of polymers as coagulant aides in the water treatment plant caused sticky sludges, which plugged the scrubber and ribbon screw. Similarly, the use of sodium aluminate caused scaling in the reactor.
11. Cost-saving conversion to a selective lime process in the water treatment plant eliminated an unrecognized major source of soda ash to the calciner. Dust buildup, uncontrolled particle size and eventual calciner failure resulted.

As these problems surfaced, immediate attention was given to the more urgent and less costly items. Forced ventilation of the calciner using a roof-mounted exhaust fan improved operator comfort. A plaster wall partition was installed to isolate the chemical storage and reduce dust problems. In 1972, and again in 1975, the sludge lagoon dikes were improved and elevated to provide additional storage for untreated wastes. A spare impeller was purchased for the exhaust fan and a sandblasting program initiated to minimize fan imbalance. In 1978 the fan was remounted on vibration arrestors to protect against further structural damage to the building. The use of polymers and sodium aluminate in the water treatment plant was discontinued. In 1979 the staff contrived a method of solid soda ash feed to overcome the problems introduced by selective lime treatment.

Detailed engineering study also was continued. A trailer-mounted pressure filter system was leased for approximately three weeks. Tests were made on various types of sludges, both before and after the calciner process. Similar tests were made using a leased centrifuge. Visits were made to Dayton, Ohio to examine sludge recarbonation facilities and discuss its recarbonation test program. The Atlanta, Georgia water treatment plant pressure filter system was visited. At the same time, a tour was made of a pressure filter manufacturing facility in Birmingham, Alabama.

These studies and plant visits led to the 1974-75 construction of a sludge recarbonation basin. The basin was placed between the calciner scrubber and the sludge thickening tank. Exhaust gases were captured prior to discharge to the atmosphere and recirculated to the recarbonation basin. the intended goals were accomplished. The process acted to dissolve the magnesium hydroxide sludges, resulting in increased settleability in the thickener and improved consolidation in the centrifuges.

For those considering such a process, there are side effects that must be evaluated. By dissolving the magnesium hydroxide, entrapped color is released. The color carries over in the thickener overflow preventing recirculation back to the water treatment plant influent. Severe scaling can result from the unstable dissolved magnesium carbonate trihydrate. This can add significantly to maintenance costs and reduce life of downstream sludge piping. Removal of the gelatinous magnesium hydroxide also affects sludge particle size, requiring consideration in downstream processes. Frequent cleaning and maintenance of the blower and gas spargers provide yet another headache.

The improvements mentioned above were accomplished at a cost of approximately $250,000 over a period of six years. Despite the improved calciner production resulting from the recarbonation basin construction, it became apparent that the shortage of thickener capacity, the lack of reliable calciner backup and the nagging demands of those "minor" residual wastes dictated the need for Ann Arbor to implement another major construction program.

In 1979 bids were received for the construction of a second thickener tank, the purchase of pressure filter equipment and the construction of a new pressure filter building. The thickener was to be available for service in the summer of 1980, with the pressure filter startup scheduled for the spring of 1981. The total cost of these improvements was expected to be $3.4 million.

Thickener capacity will be doubled, allowing for 24-hour continuous operation of the calciner. Alternate shutdown of the thickeners will now be possible, allowing for periodic inspection and maintenance. Two pressure filters are to be installed, each having adequate capacity to handle the total calciner waste. Together, the two units can adequately treat the raw sludge during periods when the calciner is out of service. Additional filter frame space was provided to allow for future expansion to meet the increased sludge loadings in future years, as the water treatment plant approaches its design capacity (fortunately, attractive pricing allowed for purchase of the additional filter plates under the current contract). A final backup to both the calciner and the pressure filter plants will continue to be provided by the existing sludge lagoons. The discharge permit for the lagoons is presently being renewed.

Because of the negative impact on particle size resulting from recarbonation, plans call for this process to be discontinued. It is expected that the carbon dioxide can be diverted to treated water recarbonation, allowing for salvage of the blower system and a reduction in the fuel needs for water treatment. The dried cake (approximately 50% solids) from the filter process is to be trucked to the sanitary landfill. Indications are that the cake will be suitable for the intermediate cover at the landfill, reducing the city's need to purchase clay for this purpose. Although unlikely, agricultural utilization may provide an alternative for disposal of the cake.

So, again, Ann Arbor has taken "wise and significant" steps. And if it were not for the potential replacement of the city's landfill with a shredder and incinerator, the limit on the sweetening of farmland with lime sludges, the fairly demanding measure of zero discharge and that this is a constantly changing environment, it might be concluded that the task had been completed.

REFERENCES

1. Scott, J. C. "Ann Arbor's Recalcining Process and Problems," *J. Am. Water Works Assoc.* 61(6):285-288 (1969).

SECTION IV

MONITORING THE ENVIRONMENT

CHAPTER 15

ENVIRONMENTAL IMPACTS OF
SLUDGE DISPOSAL

Raymond C. Loehr

> Environmental Studies Program
> College of Agriculture and Life Sciences
> Cornell University
> Ithaca, New York

The development of environmentally sound methods of wastewater treatment and sludge disposal has been a continual challenge. The treatment of wastewaters not only produces purified effluent, but also a significant quantity of sludge. Each year nearly six million dry tons of sewage sludge are generated by wastewater treatment plants in the United States. The national objective to restore and maintain the chemical, physical and biological integrity of our surface waters and groundwaters is requiring secondary and, in some locations, tertiary treatment of wastewater. Meeting these objectives will result in yet greater production of sludge.

Sludge generated from plants treating domestic sewage is essentially organic, although measurable quantities of metals, minerals and other compounds invariably are present. The sludge also may contain pathogenic organisms that survive the wastewater treatment processes. Where the wastewaters contain industrial and commercial wastes, the potential for toxic materials in the sludge is increased.

The disposal of sludge is a complex problem that can affect the air, land and water. It requires consideration of human and animal health, plant growth and nuisances, as well as protection of groundwater and surface water. Proper operation, maintenance and monitoring of sludge utilization or disposal practices is essential to avoid adverse environmental impacts.

Reflecting the public interest in environmentally sound sludge management approaches, sludge disposal is subject to legislative and social constraints: federal and state laws and regulations on disposal and reuse; public and private incentives for conventional and alternative wastewater treatment processes; and social attitudes toward waste disposal and reuse. Federal legislation affecting sludge management and disposal is noted in Table I, and the impact of such legislation on sludge management is shown in Table II.

Sludge is the inevitable product of our standard of living and our desire for a better environment. Sludge must be considered as a potential threat to the environment (negative impact) and as a potential resource that should be utilized (positive impact). Environmentally sound and economic methods of sludge disposal must be utilized. This chapter focuses primarily on the adverse environmental impacts of various methods of sludge disposal because these are of the greatest concern to the public.

Table I. Recent Environmental Legislation Affecting Sludge Management [1]

· Federal Water Pollution Control Act Amendments - 1972
· Marine Protection, Research and Sanctuaries Act - 1972
· Toxic Substances Control Act - 1976
· Resource Conservation and Recovery Act - 1976
· Clean Water Act - 1977

Table 2. Impact of Legislation on Sludge Management

· Cleaner effluents and greater quantities of sludge
· An end to ocean disposal of sludge by December 31, 1981
· Emphasis on recycling of sludge
· Greater consideration of land application of sludge
· Prevention of toxic levels of materials from entering the environment
· Industrial waste pretreatment to reduce toxic compounds in sludge
· Cradle-to-grave control of toxic and hazardous wastes

SLUDGE MANAGEMENT ALTERNATIVES

Alternative Decisions

Sludge disposal decisions are related to overall sludge management decisions, which include sludge generation, handling, stabilization, transport and disposal (Figure 1). In the ideal case, the consulting engineer, public works manager or treatment plant superintendent chooses among a number of alternatives, for which there are a number of outcomes. In practice, the number of alternatives is reasonably narrow, with many constrained by pollution control regulations, siting problems and economics.

At most wastewater treatment plants the major sludge management decisions relate to sludge processing and disposal alternatives. It is important to recognize, however, that decisions about sludge disposal alternatives are not isolated, but rather are a function of the characteristics of the sludge, which are influenced by the constituents of the wastewater and the sludge processing methods. Thus, any changes in wastewater constituents (such as by exclusion of toxic and hazardous wastewater components) or sludge processing (such as technologies to reduce pathogen concentrations, enhance nutrient concentrations or remove metals or toxic chemicals) will affect sludge disposal options and the environmental impact of those options.

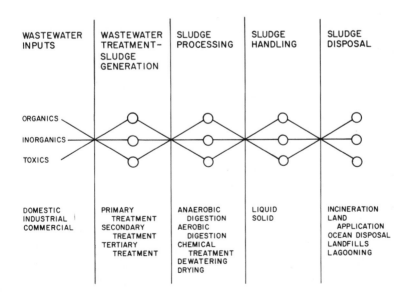

Figure 1. The many sludge management decisions that affect sludge disposal options.

Relative Risk

In considering environmental impacts of sludge disposal, it is important to consider the relative risks associated with sludge disposal options. Each option will involve environmental risks and there can be no hope of reducing all risks to zero. Because safety is a judgmental decision, a disposal option such as land application or incineration will be judged safe if its risks are judged to be acceptable by the public. Whether something is "safe" cannot be measured because the physical and biological sciences can assess only the probabilities and consequences of events, not their value to people.

The many concerns about acceptable risk can be summarized by several basic considerations, which should be taken into account when judging the severity of the environmental impacts of sludge disposal methods:

1. definition of conditions of exposure, i.e., who or what will be exposed, to what, in what way and for how long;
2. identification of adverse effects, i.e., what is the threat or adverse effect to the individuals or objects that are exposed;
3. relationship of exposure with effect, i.e., how much adverse effect results from how much exposure;
4. is the risk assumed voluntarily or involuntarily; and
5. whether the consequences are reversible or irreversible?

Sludge Disposal Options

There are many sludge disposal options, some of which are noted in Figure 1. Those considered here are: high-temperature processes (incineration, cocombustion, wet air oxidation), ocean disposal, land application (composting, agricultural land, landfill, dedicated land) and lagooning. Each of these options will be described briefly and the primary and secondary environmental impacts discussed in detail.

ENVIRONMENTAL IMPACTS

Ocean Disposal

Ocean disposal has followed one of two alternatives: (1) containment of the solids in a relatively small area of ocean bottom; or (2) disposal over vast areas. The techniques used include barging and release at sea, discharge through sludge outfalls, and disposal into wastewater effluent outfalls.

The U. S. Environmental Protection Agency (EPA) has issued criteria to govern the disposal of wastes to the marine environment. The criteria establish acceptable levels for constituents contained in the material to be disposed. Requirements for mercury and cadmium have been established at 0.75 and 0.6 mg/kg, respectively. Even domestic sludges are not likely to meet these criteria, so ocean disposal of sludge will be phased out by 1981. Coastal cities that have been discharging sludge to the ocean have begun to implement alternative disposal strategies.

Although as of 1975 about 15% of the sludge produced in the United States was disposed of in the ocean, this approach has been employed by comparatively few municipalities, such as New York City, Philadelphia, Los Angeles and a few other large coastal cities.

The environmental concern about ocean disposal focuses on the items noted in Table III. The severity of the potential environmental impacts is dependent on the constituents of the sludge, ocean currents, biological, chemical and physical processes affecting sludge stabilization, and mixing within the ocean depths.

Oil, grease and sewage artifacts can remain on or near the ocean surface after sludge disposal. These materials may discourage recreational use of the waters and may wash up on bathing beaches. In coastal areas, however, more floating material may be contributed by rivers, stormwater runoff and vessels than by ocean sludge disposal.

Although most sludges are stabilized prior to ocean disposal, they will still exert an oxygen demand, which can contribute to low dissolved oxygen concentrations in areas where sludge solids are concentrated. Plankton supported by nutrients derived from sludge and other sources can also reduce dissolved oxygen concentrations in coastal waters.

Pathogenic bacteria and viruses that exist in the sludge can be transferred to humans through ingestion of contaminated shellfish and water. They also can be transferred by body contact recreational activities, such as swimming and boating.

Table III. Environmental Impacts of Ocean Disposal

- · Floating debris and surface films
- · Depletion of dissolved oxygen concentrations
- · Alteration of benthic communities
- · Pathogens
- · Toxic compounds such as heavy metals,
     pesticides and chlorinated organics
- · Excessive increase in nutrients

Certain toxic chemicals, such as chlorinated organics (DDT and PCB), are lipid soluble. They tend to be concentrated in the lipid fraction of fish and shellfish and can be transmitted through human and bird food chains when such aquatic life are eaten. The majority of heavy metals are insoluble as sulfides, hydroxides or carbonates. They will be precipitated with the sludge solids and will increase the metal content of bottom sediments.

Where large quantities of sludge solids accumulate they can modify benthic communities. The diversity and abundance of benthic fauna has been modified in the New York bight area, where New York City sludge has been disposed, and on the Middle Atlantic Continental shelf, where there has been disposal of sludges from Philadelphia and Camden, New Jersey. High concentrations of metals can harm marine organisms, and the constituents in sludge can cause diseases such as shell erosion in crustacea, fin erosion in fishes, or abnormal growth patterns of fish and other aquatic life.

Predictions about the environmental impact of ocean disposal of sludge are tenuous. There is inadequate information on the dispersal of sludge solids; rates of decomposition; the nature of benthic fauna in areas receiving sludge and in nonsludged areas; uptake of toxics and transport of pathogens; and production of aquatic life in the offshore sludged and nonsludged regions. It is clear that the environmental effects of ocean sludge disposal are not easily controlled after the sludge is disposed of in the ocean.

High-Temperature Processes

High-temperature processes such as pyrolysis, incineration, cocombustion and wet air oxidation are not total sludge disposal methods because there always is a residual that requires further treatment or disposal. These residuals can be ash, incompletely oxidized liquids, and solids and liquids from air pollution control devices.

Pyrolysis

Pyrolysis is the high-temperature decomposition of organics in the absence of oxygen. Pyrolysis generates three forms of residuals: (1) solids such as fixed carbon; (2) liquids such as methyl alcohol, acetones, oils, residual tars and water; and (3) gases such as carbon monoxide, carbon dioxide, hydrogen and methane.

The yield and composition of the residues depend on many variables. The actual interrelationships are complex, and final product characteristics must be determined empirically. As of 1979 there were no full-scale pyrolysis projects under development that used sludge alone. Most of the available information has originated from laboratory studies.

Incineration

Incineration is a two-step oxidation process involving drying and combustion. The drying and combustion may be accomplished in separate units or successively in the same unit. Prior to incineration, the sludge is dewatered mechanically. The remaining water is removed in the incinerator drying step. In all furnaces, the following steps are involved: (1) raising the temperature of the feed sludge to 100°C (212°F); (2) evaporating the water from the sludge; (3) increasing the temperature of the water vapor and the air; and (4) increasing the temperature of the dried sludge volatiles to the ignition point. Incineration is a complex process involving thermal and chemical reactions that occur at varying times, temperatures and locations in the furnace.

The purposes of incineration are volume reduction and sterilization of the end product. Depending on the characteristics of the sludge, about 10-30% of the initial weight of dry solids will remain as ash. The volume of the dewatered sludge will be reduced to a much greater degree.

Cocombustion

Direct combustion of sludge is possible only when the sludge solids content is 30-35% or greater. This content is difficult to achieve by conventional dewatering methods, and supplemental fuel is required for the combustion. If sludge is combined with other combustible material in a cocombustion process, a furnace input can occur having both a low water concentration and a heat value high enough to sustain autogenous combustion.

Many materials can be combined with sewage sludge for cocombustion. These include coal, municipal solid wastes, wood wastes, sawdust and crop residues such as corn stalks, rice husks and bagasse. An advantage of cocombustion is that municipal or industrial waste often can be disposed of while providing an energy self-sufficient sludge-waste feed, thereby solving two disposal problems.

The residuals resulting from cocombustion are the same as those from incineration, i.e., gases and ash. Cocombustion is a relatively new venture. Its use must be evaluated and tested thoroughly, and project economics must be identified clearly.

Wet Air Oxidation

Wet air oxidation is similar to other thermal processes except that more air, lower temperatures and higher pressures are used to obtain oxidation. The process can be operated in various modes, depending on the desired goal or end product. The degree to which organic materials are oxidized is a function of reaction time, quantity of oxygen supplied, temperature and pressure. Up to about 50% total oxidation, reduction of the volatile solids or chemical oxygen demand (COD) in the liquid phase is minimal. At 80% total oxidation, about 5% of the original total solids in the sludge is in the solid phase and about 15% in the liquid phase.

Wet air oxidation may be applied to thickened sludge. The solids content should be 4–6% to minimize reactor volume requirements and maintain a thermally self-sustaining reaction. A source of high-pressure steam must be provided for startup.

The main residues from this process are the oxidized sterile solids, which can be dewatered for final disposal; a liquid residual from dewatering, which is recycled to the wastewater treatment plant; gases that are processed by wet scrubbing, adsorption, afterburners or catalytic oxidation to remove odors, sulfur and nitrogen oxides, incompletely oxidized hydrocarbons, and particulates.

Environmental Impact

The environmental concerns related to high-temperature sludge disposal relate to the items listed in Table IV.

The ash resulting from high-temperature processes can be either dry or contained in scrubber water. The common method of ultimate ash disposal is disposal on land, either in ash lagoons or in landfills. Ash disposal approaches must be designed to protect groundwater, minimize dust production and ensure protection of surface waters. Surface runoff from ash disposal sites should be avoided, and extraneous, upstream runoff diverted around such sites.

High-temperature sludge disposal processes can contribute to air pollution because of incomplete combustion and formation of intermediate combustion products. All such units must be equipped with air pollution control equipment to meet air quality standards.

**Table IV. Environmental Concerns Related to
High-Temperature Sludge Disposal Methods [2]**

· Disposal of ash
· Air pollutants
  Organics
  Metals
  Particulates
  NO_x, SO_x
· Odor

The intimate mixing of air with sludge in incinerators provides the opportunity for the noncombustible fraction of sludge to be carried away by the exhaust gases. Thus, incinerators emit fine-grained fly ash. To meet particulate emission standards, high-energy scrubbers, baghouses or electrostatic precipitators are needed because simple water sprays or baffle/setting chambers are inadequate. Scrubbers generate a liquid waste that must be handled and treated in an environmentally sound manner.

The nitrogen and sulfur content of sludge can form nitrogen oxides and sulfur oxides during high-temperature combustion. Research has indicated that sludge incinerators are a minor source of nitrogen oxides.

Sludge incinerator emissions can contain PCB, a volatilized mercury and particulates containing trace amounts of metals such as lead or cadmium. If sludge contains high concentrations of PCB (generally above 25 mg/kg dry sludge), incinerators should be operated at temperatures that will ensure destruction of such compounds. Generally, 95% destruction of PCB is achieved in a multiple-hearth sludge incinerator with no afterburning at an exhaust gas temperature of 370°C (700°F). At 870-980°C (1600-1800°F), there is 99% destruction in two seconds. Increased temperature and incinerator sludge residence time increase PCB destruction.

High temperatures cause volatilization and emission of mercury, arsenic, cadmium and lead. The impact of these emissions on the environment appears to be small because the amounts in sludge are small, as is the fraction emitted to the atmosphere. Limited data indicate that about only 4–35% of the mercury entering an incinerator with emission controls would be emitted in nonparticulate form to the atmostphere. Wet oxidation emits no fly ash but may give off small amounts of organic, odorous gases. Incomplete combustion in incinerators can result in odors and increased organic and particulate emissions.

Land Application

The land application systems that can be used for sludge disposal include composting followed by land application of the compost, landfills, application to parks and similar lands, application to agricultural land, and application to dedicated land sites.

Composting

Sludge composting is the aerobic thermophilic decomposition of organic constituents to a relatively stable, humus-like material. The factors that influence composting are the moisture content, mixing and availability of oxygen, carbon–nitrogen ratio, temperature, pH and the type of the material. The composting process is considered complete when the product can be stored without generating odors and when the pathogenic organisms have been reduced significantly.

Compost produced from municipal sludges can provide nutrients for crops and is beneficial as a soil conditioner. The improved physical properties of soil that can result from addition of compost include: increased water content and water retention for sandy soils; increased aeration, permeability and water infiltration for clay soils; and decreased surface crusting. The composting process results in a significant nitrogen reduction in the sludge and, therefore, a reduced amount of nitrogen available to soil and plants.

Sludge composting methods include unconfined processes, such as windrow and aerated static pile composting, and confined processes. Unconfined processes make use of portable mechanical equipment for mixing, turning and aeration. Confined systems utilize an enclosed reactor for composting.

Landfills

A landfill for sludge disposal involves the planned burial at a designated site of processed sludge and other residues generated at a wastewater treatment plant, such as screenings, grit and ash. The solids are placed into a prepared hole or excavated trench and covered with a layer of soil. The soil cover must be deeper than the plow layer. Sludge landfills are grouped according to whether they are sludge-only trench landfills, sludge-only area landfills and codisposal with refuse landfills.

Narrow-trench landfills are more applicable to liquid sludge disposal. Generally, the sludge is stabilized and dewatered before being landfilled. Liners could be needed where the groundwater occurs at a shallow depth.

Agricultural Land

Liquid or dewatered stabilized sludge is applied to land as a method of final disposal; of land reclamation; and as a source of nutrients for grasses and other crops. Because land application conserves and recycles organic matter, nitrogen, phosphorus and trace elements, EPA encourages such utilization when it is supported by an environmental assessment and is well managed.

Liquid and dewatered sludges are applied by landspreading using tank trucks, sprinkler systems designed to handle solids without clogging, and open trucks. The components of a sludge landspreading system are sludge processing, transport, storage, land application and incorporation. Each component influences the selection of the proper design for the other components.

A number of factors influence the design and operation of a land application system: (1) the acreage needed, which is determined by the parameter limiting the application rate; (2) sludge and soil characteristics; (3) climate; and (4) the type of crop to be grown. Soil characteristics that often are the most important factors in site selection include pH, electrical conductivity, organic matter content, natural nutrient level, cation exchange capacity, natural level of metals and structure.

Dedicated Land Disposal

Dedicated land disposal means the application of heavy sludge loadings to some finite land area that has limited public access and has been set aside or dedicated for all time to the disposal of wastewater sludge. As with any other land disposal method, dedicated land disposal requires the sludge to be stabilized prior to application.

The use of dedicated land disposal has the following advantages:

1. flexibility to manage sludges in excess of agricultural land utilization rates;
2. smaller land use because application rates per acre are higher; and
3. lower capital and operating costs.

Items to be considered when selecting appropriate dedicated land disposal sites include present and future ownership, groundwater patterns, topography, soil types and availability of sufficient land.

Environmental Impact

The important environmental concerns related to land application are noted in Table V. At composting sites, the major environmental concerns relate to pathogen control, odors, dust and runoff from the site [2]. Extensive studies have shown that if the temperature in the compost reaches

Table V. Environmental Concerns Related to the Land Application of Sludge

Composting	Landfills	Land Application	Dedicated Land
Pathogens	Odors	Food chain	Leachate
Odors	Rodents and flies	Metals	Odors
Dust	Erosion and runoff	Toxic organics	Erosion and runoff
Runoff	Leachate	Odors	Pathogens
	Aesthetics	Erosion and runoff	
		Leachate	
		Pathogens	

the 50-60°C range, fecal and total coliforms, salmonella and most viruses are destroyed in 10-15 days in windrow or static pile systems. Very low levels of parasitic ova, virus and other pathogens have been measured in most final compost samples.

Data have shown that high concentrations of the fungus *Aspergillus fumigatus* can be airborne at composting sites. Generally, the high concentrations are restricted to the immediate composting area and should not pose a significant health threat to surrounding residential, commerical or industrial areas. Respiratory protection, such as breathing masks, are advised for workers at the site to avoid inhaling dust, fungal spores and other airborne irritants.

A drainage collection system is needed for stormwater runoff at the site to avoid pollution caused by site-contaminated runoff. The runoff may be returned to the wastewater treatment plant or irrigated onto adjacent land.

Odor control is a prime environmental consideration in the operation of composting systems. Poor mixing and high rainfall can lead to anaerobic conditions and related odors. The compost should be adequately cured before it is removed, and unstabilized material should be recycled for further treatment. Air from an aerated pile composting system should be discharged through an odor filter pile made up of stabilized compost.

The major environmental impacts from landfills include odors, pollution of surface and groundwaters, unsightliness and proliferation of rodents and flies [2]. Covering the landfill daily with soil can control odor and disease vector problems. Equipment noise, dust and spillage can be lessened by barrier vegetation and good site management. While improper siting and excavation of landfills can increase erosion, gradient limitations, drainage diversion, vegetative barrier strips and sedimentation basins can control erosion problems.

Leachate from a landfill may transport nitrate, metals, toxic organics and pathogens to groundwaters, but both clay and plastic liners and proper site selection can reduce and avoid such problems. A properly sited, designed, constructed, operated and managed sludge landfill is an environmentally safe method for sludge disposal.

The primary environmental concerns related to land application of sludge relate to potential contamination of the food chain and pollution of surface and growndwaters [2]. In addition, there can be public health, odor and nuisance problems. Sludge applications on the land can generate serious odors if the site and application rates are not properly managed. Odor problems begin at the point of initial sludge handling and continue after the sludge is applied to the land. The degree of offensive odor depends on the type and nature of the sludge, any pretreatment or dewatering prior to disposal, and how it is managed after it is applied.

The sludge should not be allowed to stand in liquid pools for any length of time and should be incorporated into the soil when applied or shortly after. Tank trucks transporting the sludge to the application site should be clean and leakproof.

The potential for pathogen transmission exists and can cause a public health problem if the land application is done improperly. The transmission can occur through groundwater, surface runoff, aerosols formed during application and direct contact with the sludge or raw crops from the application site.

Because bacteria, viruses and parasites do not enter plant tissue, transmittal of pathogens via crops grown on the land application site results from contamination of the plant surface. If contaminated crops are consumed raw, disease transmission is possible; however, disease transmission due to application of sludge onto farmland is rare. Reported outbreaks of disease generally have been the result of application of inadequately treated sludges to truck gardens or other crops which were eaten raw.

Pathogens in land-applied sludge usually will die rapidly depending on temperature, moisture and exposure to ultraviolet light. Typical survival times in soil and on plants are noted in Table VI. In general, pathogen survival is shorter on plant surfaces than in the soil. To prevent disease transmission, sludge should not be applied to land during a year when crops are to be grown that will be eaten. Where humans have little physical contact, the presence of pathogens may be of less concern. The soil can filter and inactivate bacteria and viruses. Sludge application methods and rates should take advantage of the soil to reduce public health concerns.

Another potential constraint is the possibility of increased nitrate concentrations in the groundwater and transmission of heavy metals and toxic

Table VI. Survival of Certain Pathogens in Soil and on Plants

Organism	Media	Estimated Survival Time (months)
Bacteria	Soil	Up to 6
	Vegetation	0 to 3
Enteric Viruses	Soil and Vegetation	0 to 3
Protozoa	Soil	Up to 6
	Vegetation	0 to 2
Parasites (ova)	Soil	Up to several years
	Vegetation	1 to 2

organics through the food chain. Certain metals also are known to be toxic to specific crops. Most states have guidelines and regulations controlling the quantity of metals and toxics that should be applied to land. In addition, EPA has promulgated criteria for solid waste disposal facilities and practices. These include criteria for the application of sludge to land used for the production of food chain crops [4].

Two interrelated key factors in avoiding adverse environmental impacts from sludge land application systems are the sludge application rate and the land area that is used. Many factors determine the required land area, such as sludge characteristics, characteristics of the soil, climate, wastewater and crop. These should be evaluated using site-specific information.

The application rate of the following parameters will significantly affect the required land area:

1. water,
2. organics,
3. nutrients,
4. potentially toxic elements, and
5. salts.

When obtaining the required land area, the land area for each potentially limiting parameter should be determined. That parameter requiring the largest land area to avoid environmental problems becomes the limiting parameter. This "limiting parameter principle" states that the design land area shall be no less than that allowed by the limiting environmental parameter.

Figure 2 illustrates the concept of the limiting design parameter. In the example, nitrogen is the controlling design parameter, as is the case in most land application systems for disposal of municipal sludge.

When the land area determined for the limiting parameter is used, there is an added degree of safety in terms of the application rates of the other constituents of potential concern. The application rate of another constituent will be considerably less than the rate that would occur if the constituent were the limiting parameter.

The basic concept inherent in the limiting parameter approach is to use site-specific data to meet the desired groundwater quality and calculate the loadings accordingly. The concept uses soil loading criteria, which incorporate specific information about the sludge characteristics, soil characteristics and the crop for the design of an environmentally sound land application system for sludges.

Although there can be adverse public health, food chain and groundwater contamination problems, land application of sludge can be a practicable method of sludge disposal provided that the system is carefully, efficiently and continuously managed; crops are restricted to those not eaten raw; and

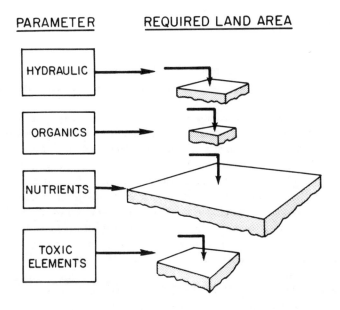

Figure 2. An example of the limiting parameter principle to determine the required land area for land application and to avoid adverse environmental impacts.

monitoring exists to detect and control potential public health threats. It also has been noted that to date there is no epidemiological evidence to suggest that land application of sludge has resulted in actual human illness where sludge has been properly treated and applied [5,6]. Further, because of marketing practices and types of crops commonly grown on sludge-amended soils, health effects from cadmium in municipal sludge are not expected to be a problem.

The environmental impacts of dedicated land disposal of sludge are similar to those of the application of sludge to agricultural or nonagricultural land with the following major exceptions:

1. No food chain crops should be grown on dedicated land disposal sites;
2. There can be a significant accumulation of metals and nonbiodegradable organics in the dedicated land site [2].

The limiting parameter principle for dedicated land disposal sites would not include crop uptake in mass balances for nitrogen and metals and may not be concerned with application rates of heavy metals except to ensure that excess quantities do not leach to the groundwater.

Lagooning

Lagooning in excavated or natural depressions can be a feasible sludge disposal method if land is readily available and inexpensive. Lagooning may be considered as a part of the handling of sludge or as a final disposal process.

Sludge solids settle and accumulate in a lagoon, and any excess liquid is returned to the plant for treatment. If the lagoon is used only for digesting sludge, nuisance problems such as odors are minimal. Lagoon disposal should be limited to digested or stabilized sludges to eliminate possible odor and insect problems. Where adequate land is available, sludge lagoons may be large enough that the settled sludge solids are never removed, or at least not removed for many years. The sludge solids that are removed must be disposed of in a landfill or on available land using properly designed and managed approaches. In this sense, lagooning is not an ultimate disposal method.

The environmental impacts related to sludge lagooning concern groundwater contamination by leachates as well as odor and insect breeding. To minimize leaching, a lagoon site should be at a site with an impervious soil and a low water table. Surface runoff should be diverted around the lagoon.

SOCIETAL IMPACTS

Much of the available information on sludge disposal alternatives focuses on technical, health or ecological aspects. Public acceptance is equally important to the success of such alternatives. For many metropolitan areas, sludge disposal can involve some form of export to sites in low population areas. Even if such sites are located in the same political jurisdiction, local opposition to the acceptance of sludge from someplace else is often intense and can escalate considerably when the sludge is from another political jurisdiction. Opposition can be so great as to preclude certain sludge disposal options.

Public concerns focus on the legality, public health and nuisance aspects of the sludge disposal option. Many options remind people of situations in which improperly handled sludge caused odors, surface and groundwater pollution, nuisances to those living nearby and reduced property values. The social costs related to health hazards, insults to aesthetic sensibilities, land lost from tax roles and feared reduction of property values are difficult to quantify, yet real to those who feel they are impacted.

Because sludge disposal options ultimately are located within the boundaries of some local government, the regulations and laws of that local government as well as those of the state and federal government must be considered and utilized to avoid adverse technical, ecological and societal environmental impacts. Regulations and laws that affect land use, public health, water pollution, air pollution and nuisances must be complied with.

Minimizing adverse technical or ecological environmental impacts can be achieved by adhering to existing regulations, laws, ordinances and guidelines. An even more important factor for the success of a sludge disposal option is adhering to sound operating and management procedures. Even the best designed and located sludge disposal option may be unsatisfactory because of inadequate operation and management. However, even with good design, location, operation and maintenance, the perception of the local residents may be the governing factor in the choice of a sludge management alternative.

SUMMARY

Each of the many sludge disposal alternatives has potential environmental impacts relating to one or more of the following: groundwater and surface water pollution, odors, contamination of food chain crops, transmittal of pathogens, air pollution, ash disposal, alteration of terrestrial and benthic communities and nuisances. Often there is a perception that these impacts are greater than they are, especially to individuals living adjacent to sludge disposal sites.

The fact that there can be a number of adverse environmental impacts is less relevant than the relative risk associated with the impacts. The factors noted earlier, i.e., length and amount of exposure, individuals or objects exposed, and the reversibility of the impact can help identify the relative risks of obvious sludge disposal options.

Almost all the potential environmental impacts can be controlled by sludge disposal options that are well designed, carefully, efficiently and continuously managed, and that follow accepted guidelines and regulations.

REFERENCES

1. *Multimedium Management of Municipal Sludge* - Volume IX (Washington, D. C.: National Academy of Sciences, 1977).
2. U. S. Environmental Protection Agency. *Process Design Manual for Sludge Treatment and Disposal,* EPA 625/1-79-011, Office of Technology Transfer, Washington, D. C. (1979).
3. LA/OMA Project. "Sludge Processing and Disposal - A State of the Art Review," Regional Wastewater Solids Management Program, Los Angeles/Orange County Metropolitan Area (1977).

4. "Criteria for Classification of Solid Waste Disposal Facilties and Practices; Final, Interim Final, and Proposed Regulations," *Federal Register* 52438-53464 (September 18, 1979).
5. Pahren, H. R., J. B. Lucas, J. A. Ryan and G. K. Dotson. "Health Risks Associated with Land Application of Municipal Sludge," *J. Water Poll. Control Fed.* 2588-2601 (1979).
6. Wolman, A. "Public Health Aspects of Land Utilization of Wastewater Effluents and Sludges," *J. Water Poll. Control Fed.* 2211-2218 (1977).

CHAPTER 16

THE IMPACT OF SLUDGE INCINERATION
ON THE ENVIRONMENT

Walter R. Niessen
>   Camp Dresser & McKee Inc.
>   Boston, Massachusetts

For many, there is an instinctive negative emotional reaction to the concept of incineration. When applied to sewage sludge, the reaction is further reinforced. This chapter assesses the impact of sludge incineration on an objective basis and presents a basis for evaluation that allows fair comparison among the numerous alternatives available to the sludge management specialist.

Table I compares the five most common sludge management methods in various impact catagories. The air pollution impacts ascribed to land application, composting and landfill include the vehicular emission associated with the hauling operations, as well as the particulate and odor that may arise during and after field application.

The question marks shown in Table I indicate an important characteristic of the nonincineration methods relative to the assessment of their impact: many of the impacts are not quantitatively known, although they are not necessarily insignificant. For incineration, however, the characteristics of the process are such that the various process streams can be readily sampled and analyzed and the impacts quantitatively reported. This tends to make incineration "look bad" when an objective assessment backed with sound data might show incineration to have a comparable or even lesser impact than its "competitors."

Table I. Environmental Impact Spread Sheet

Impact Category	Incineration	Land Application	Composting	Stabilization and Landfill	Ocean Disposal
Air Pollution					
Odor	—	Moderate	Moderate	Low	—
Particulate	Low	?	?	Very low	—
CO, NO$_X$	Low - Medium	Varies	Varies	Varies	Very low
Heavy metals	Low	?	?	—	—
Water Pollution					
Surface water	—	Varies	Varies	—	—
Groundwater	Very low	Moderate	Moderate	Low	—
Land Use	Very low	Moderate - high	Moderate - high	Low	—
Traffic	Very low	Moderate	Moderate	Low - moderate	—
Energy	Gain - high	Very low - moderate	Low - moderate	Low - moderate [a]	Very low
Health (pathogen)	—	?	?	—	?
Biota	—	?	?	—	?

[a] Could be some offsetting energy gain if digestion is used for stabilization.

Figure 1 schematically indicates the major environmental impacts associated with sludge incineration. These include air pollution emissions; water pollution emissions from the scrubber water and boiler blowdown; commitment of land resources for ash landfills; traffic impacts in the hauling of the ash; groundwater impacts from residue disposal; and energy impacts, which may be beneficial or adverse depending on the overall energy balance in the incineration system.

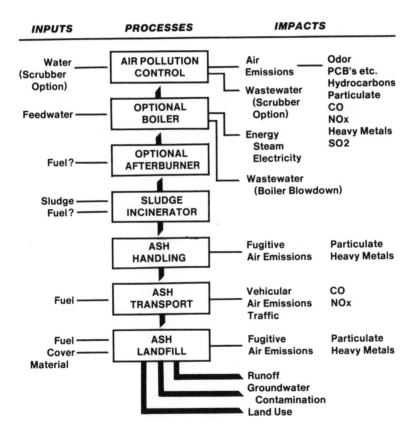

Figure 1. Environmental impacts of sludge incineration.

AIR POLLUTION IMPACTS OF SLUDGE INCINERATION

Table II summarizes the air pollution characteristics of the sludge combustion process. The pollutant sources on the left side correspond to the pollutant emissions shown on the right.

Air Emissions

The sludge volatile content, approximately equivalent to the combustible content, refers to that portion of the sludge that is normally burned off in the incinerator. If, however, the incineration conditions near the exit flue are inadequate to fully burn the volatiles, these pollutants may be emitted in the gas stream, leaving the incinerator relatively unchanged chemically from the form in which they appear in the raw sludge [1]. These pollutants can include the polychlorinated biphenyls (PCB), various odorous compounds and various low-molecular-weight hydrocarbons.

The sludge ash constitutes the majority of the particulate material leaving the incinerator and includes the heavy metals, which are increasingly a matter of concern to health and regulatory officials.

Table II. Air Pollution

Pollutant Source	Notes	Pollutant
Sludge Volatile	a,b	PCB, etc.
	b	Odor
	b	Hydrocarbons (HC)
Sludge "Ash"	—	Particulate
	a,c	Heavy Metals
Combustion Process	b	Carbon monoxide (CO)
	b	Partially oxidized HC
	a,d	Sulfur Oxides (SO_2, SO_3)
	e	Nitrogen oxides (NO_x)
Ash Handling and Landfill	a	"Ash" pollutants
Ash Transport (Vehicles)	—	CO
	—	NO_x
RDF Auxiliary Fuel		"Ash" pollutants
		Combustion process pollutants
	f	HCl

[a] Magnitude of problem depends on concentration in feed.
[b] Minimum emissions with afterburner at 1400°F.
[c] Especially cadmium, mercury, and lead.
[d] Only pyritic and organic sulfur are important.
[e] Emissions increase with peak flame temperature.
[f] Emissions arise from chlorocarbons (esp. PVC) in refuse.

The pollutants associated with the combustion process fall into two categories. The first, including carbon monoxide and partially oxidized hydrocarbons, is the result of imperfect or partial combustion, although carbon monoxide is not believed to be a significant pollutant in the effluent from sludge incinerators. Poor incineration and/or quenching of the combustion reactions can lead to the formation of odorous, partially oxidized hydrocarbons (aldehydes, ketones and more complex oxyorganics). Sulfur oxides are formed from the combustion of pyritic and organic sulfur in the sludge. It should be noted that these forms of sulfur correspond to approximately 75% of the total sulfur content reported in a typical analysis of municipal sludge. Nitrogen oxides are formed in almost all combustion reactions, both from the union of nitrogen in the combustion air with oxygen and, more especially, from nitrogen entering the combustor as nitrogenous compounds in the material being burned [1]. Generally, the low-combustion-temperature characteristics of sludge incinerators do not lead to an intolerably high nitrogen oxide emission.

Ash pollutants (arising from the suspension of particles of sludge ash) can be significant if operational practices result in fugitive dust problems in ash handling, hauling and landfill activities. Transportation of the ash to a landfill results in some minor emissions of carbon monoxide and nitrogen oxides from the vehicles.

The last item in Table II refers to pollutants arising from the use of refuse-derived fuel (RDF) to supply energy deficits in the sludge incineration process. While not common in the U. S. at this time, use of RDF as a supplementary fuel increases the amount of ash being processed in the system (compared, say, with oil or gas firing) and may result in the emissions of some hydrochloric acid (HCl) from the combustion of chloro-carbons such as polyvinylchloride (PVC) in the refuse [2].

Air Emission Standards and Controls

Table III indicates the standards on emissions that regulate sludge incinerators. The New Source Performance Standards (NSPS) and National Emission Standards for Hazardous Pollutants (NESHAPS) for sludge incinerators are currently under review by EPA. At present, it is not envisioned that these standards will be substantially changed, although some consideration has been given to modification of the NSPS limit to recognize recently perceived relationships between the minimum obtainable emission rate and the moisture content of the feed sludge.

In addition to the emission standards that apply to the sludge incinerator, the region adjacent to the treatment plant is subject to the Ambient Air

Table III. Stack Emissions Regulatory Constraints

New Source Performance Standards (NSPS)	40 CFR 60, Subpart "O" Particulate 1.3 lb/ton ds feed opacity 20%
National Emission Standards for Hazardous Pollutants (NESHAPS)	40 CFR 61, Subpart "E" Mercury 3200 g/day
Ambient Air Quality Standards (AAQS)	Particulate NO_x SO_2 Lead Carbon monoxide Cadmium (in preparation) Hydrochloric acid (in preparation)
Local Emission Regulations Local Air Quality Regulations Local Nuisance Codes (especially odor)	

Quality Standards (AAQS) for the several pollutants indicated in Table III, particularly for urban locations where the plant may be sited in an area designated as a "nonattainment area." The AAQS may require the implementation of more effective emission control systems than are commonly applied to produce emission rates less than the NSPS requirements.

Beyond the federal standards of NSPS and NESHAPS, the sludge incinerator must also be in compliance with local emission regulations, air quality regulations and nuisance codes. The latter can be an important enforcement tool if an improperly operated incineration system is plagued with odor problems. Generally, the NSPS and the NESHAPS emission rates are the basis for the assessment of impact and are incorporated into the performance specifications prepared for the incineration system and its associated air pollution control device. Thus, although different incineration systems have different *uncontrolled* emission rates, impact assessment generally revolves around the use of the NSPS/NESHAPS emission rates.

Table IV indicates the control efficiency class that may be used to broadly characterize the performance of various devices for sludge incinerator air pollution control. Clearly, the afterburner provided positive control of all combustible pollutants carried out in the flue gas leaving the incinerator, whereas only modest control of these pollutants is obtained in scrubbers and essentially none in the electrostatic precipitator. Low-pressure drop scrubbers, such as those of the cyclonic type, are relatively ineffective and

Table IV. Air Pollution Control Impact Mitigation

	1400°F Afterburner (%)	Scrubber Low ΔP (%)	Scrubber High ΔP (%)	Electrostatic Precipitator (%)
Particulate (course: >10 μ)	nil[a]	90+	100	98+
Particulate (fine: <10 μ)	nil[a]	10–20	50–95	98+
Particulate (volatile metals)	0	0–20	30–90+	?
Odorous Hydrocarbons	100	10–20	50–80	0
PCB, etc.	100	10–20	50–80	0
Hydrocarbons	100	10–20	50–80	0
Carbon Monoxide	100	0	0	0
Nitrogen Oxides	Increased	10–20	10–20	0
Sulfur Oxides	0	90+	100	0
Hydrogen Chloride	0	90+	100	0

[a] Combustible fraction burned out. Very limited data indicate 40% reduction for MHF incineration and 94% reduction for MHF pyrolysis.

generally cannot attain the NSPS level of particulate emission. High-pressure drop (10-20 inches of water) venturi scrubbers have shown ready attainment of the NSPS emission characteristics and substantial collection of fine particulate for sludges of 25% solids content.

The electrostatic precipitator has been used primarily in Europe in fluid bed incinerator applications. Most vendors are unwilling to recommend the device for multiple-hearth furnaces not equipped with an afterburner because of the buildup of tarry materials on the precipitator collection plates. However, when the particulate is well burned out (as with a properly operated fluid bed), the electrostatic preceptator provides good air pollution control with a negligible impact on the treatment plant. Scrubber water recycle can increase the hydraulic loading on the plant by as much as 5%. Also, as regulatory pressures demand still higher scrubber efficiency (pressure drop), the electrostatic precipitator readily can be shown to be more energy-efficient and cost-effective than the venturi scrubber system.

The assessment of air pollution impacts for sludge incinerators generally begins with an emission estimate developed from equipment guarantees and confirmed with stack testing in conformance with appropriate EPA test methodologies. In most assessment efforts, the system hardware is not yet developed and the NSPS levels of emissions are assumed. Subsequent air quality impact evaluations are generally made using computer diffusion modeling methods, rather than air quality monitoring. EPA–approved models, such as the Climatological Dispersion Model (CDM) and the Air Quality Display Model (AQDM), are commonly used to evaluate average annual impacts. Newer models are particularly useful in evaluating short-term impacts.

An element of design often overlooked for aesthetic reasons relates to the problems of downwash [3]. It is a well-recognized phenomenon that when the wind blows across a building a swirling eddy forms on the lee side. When the incinerator stack is short and/or the discharge velocity is very low, the emitted gases may be swept into the eddy downstream of the building, virtually eliminating all lofting of the plume due to the physically, thermal bouyancy-, or velocity-related stack height to result in a ground source of high concentration. The high concentration of pollutants in the eddy affects both the internal environment of the sludge incinerator building (if air intakes are located on the lee side of the building) and the neighborhood adjacent to the plant in the downwind direction.

In summary, a number of incineration process stages generate a variety of pollutants. These pollutants are susceptible to control with good incinerator and air pollution control system operation. Low levels of emission corresponding to the NSPS limitations are readily obtainable. Limited and unpublished data from EPA indicate that the heavy metal impacts of sludge incineration are generally unimportant relative to other sources of airborne heavy metals (e.g., lead emissions from automobiles) in the area surrounding the treatment plant. Nonetheless, heavy metal emissions deserve careful attention in the impact assessment of sludge incineration processes because data suggest that the weight fraction of heavy metals in the particulate is enriched relative to the mean concentration of the ash.

ENERGY IMPACTS OF SLUDGE INCINERATION

Table V summarizes the primary energy consumption elements of the sludge incineration process and indicates the potential generation of energy and its transformation to other useful forms (electricity or shaft power).

In the minds of many municipal officials, the high energy consumption (oil or gas) believed to be *necessarily* associated with incineration is often a major impediment to the implementation of sludge combustion alternatives. This concern over excessive fossil fuel is not always justified, although systems burning low-solids-content sludges such as the combustion of 15–20% solids/sludge (e.g., produced by vacuum filtration of waste-activated sludge) will have a high fuel demand.

In the assessment of sludge incineration, energy consumption CDM has found utility in a correlating function called the energy parameter, defined as in Equation 1:

$$\text{Energy parameter} = \frac{(1\text{-}S) \times 10^6}{(S)\ (V)\ (B)}$$

where   S = decimal % solids,
        V = decimal % volatile in solids, and
        B = Btu lb of volatile.

Table V. Energy

Energy Consumption

Electricity	Air pollution control	(especially high ΔP scrubber)
	Fans	(especially fluid beds)
Fossil Fuel	Afterburners	(especially MHF)
	Incinerators	(especially MHF, fluid bed)
	Ash transport	
	Landfill operations	

Energy Generation

Steam	(for digester heating, space heat and/or sale)
Electricity	(from steam)
Shaft Power	(from steam)

[a] In some instances, refuse-derived fuel may be substituted for fossil fuel.

The energy parameter combines the intrinsic heat content of the sludge (B), the volatile content (V) and the moisture content as reflected in the percent solids (S). The energy parameter thus expresses in a single term the balance between the latent heat requirement for evaporation of water carried in with the sludge solids to the intrinsic heat content of the sludge solids themselves. The utility of the energy parameter as an analysis taken is shown in Figures 2 and 3, where the fuel requirements and potential heat recovery for various sludges are shown. It is noteworthy that the correlations are linear and inherently account for the diluting effect of inert conditioning chemicals, the heat content differences between primary and waste-activated sludge, etc.

Referring to Figure 2, it can be seen that even for a very low-energy-parameter sludge (dry and/or high heat content), the multiple hearth furnace is *never* fully autogenous (zero fuel use). This results from limiting the top hearth temperatures of the multiple hearth to approximately 800-1000°F. Under such conditions, some fuel necessarily will be used for the afterburner even though the furnace itself becomes autogenous at an energy parameter of approximately 320.

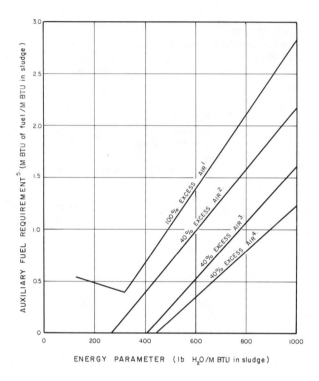

Figure 2. Fuel requirements for sludge combustion.

Figure 2 shows that for the fluid bed incinerator with a cold wind box (no regenerative air preheat) or for the pryolysis-mode multiple hearth incinerator (where the afterburner temperature is attained by flaring the hydrocarbon-rich offgases of the incinerator) there is a definite zero fuel use point corresponding to an energy parameter of approximately 260. By the use of a heat exchanger preheating combustion air with heat from the furnace offgases (the hot wind box fluid bed), the autogenous point moves to correspond to an energy parameter of approximately 400.

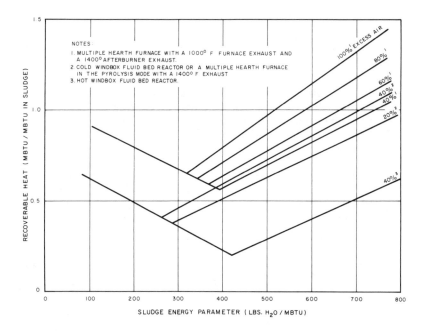

Figure 3. Potential heat recovery for sludge incineration.

With application of an afterburner, or with dewatering sufficient to generate fluid bed furnace offgas temperatures of 1500°F or more, there is a substantial amount of recoverable energy in the flue gases leaving the furnace. Figure 3 indicates the heat recoverable (as steam) in a boiler with a 450°F temperature at the cold end of the boiler.

The steam recovered from the hot furnace gases can be used for building heat or hot water, as well as for larger plants for the generation of electricity. In a recent study for the city of Detroit (1000 dry ton/day of sludge), it was shown by CDM that more than 20 MW of electricity could be generated while meeting all building heat and hot water requirements. In a plant soon to go into design at CDM for Linden, New Jersey, coincineration of sewage sludge with municipal refuse will be practiced. Steam recovered from the incineration process will be used for sludge drying and digester heating, while the remainder (>10,000 lb/hr) will be sold to an adjacent industry, thus significantly defraying the cost of incineration.

Table VI summarizes the energy use associated with transportation of sludge ash and, for comparison, the energy for transportation of sludge

Table VI. Transportation Energy Use

Basis: 100 ton/day (ds) of 40% ash sludge
20-ton payload in trucks
5 mpg fuel use in truck
20-mile equivalent round trips to disposal site
50% volatiles reduction in digestion
15% $FeCL_3$-CaO addition to attain 40% solids

Sludge Type	Ton/Day	Trucks/Day	Gallons/Year of Fuel
5% solids (raw)	2,000	100	146,000
5% solids (digested)	1,400	70	102,200
20% solids (raw)	500	25	36,500
20% solids (digested)	350	18	26,280
40% solids (raw)	287	15	21,900
Dry Ash	40	2	2,900

dewatered to various degrees. It clearly shows the superiority of sludge incineration as far as its impact on transportation energy use and indeed shows that the fossil fuel use associated with land application of thickened sludge can be substantial.

Further, it shows the traffic impacts associated with hauling of raw sludge or ash which, again, can be considerable for land application-type sludge disposal methods.

WATER POLLUTION IMPACTS OF SLUDGE INCINERATION

Within the treatment plant complex, a number of sidestreams are generated by incineration (Table VII). These include wastewater from the scrubber, boiler blowdown, and possibly cooling water, which may be used in a condenser on steam turbines.

Depending on the degree of combustion completeness in the incinerator, and/or grease and odor emissions, the scrubber wastewater will contain biochemical oxygen demand (BOD) and undoubtedly will recycle heavy metals to the plant. The continuous blowdown of a small portion of the water in the boiler system will add primarily concentrated salts (total dissolved solids) to the treatment plant, although these salts generally are present in the city water supplied to the boiler water treatment system. The salt content is augmented somewhat by the chemicals used in demineralizers and other boiler water treatment systems. There also will be some

Table VII. Water Pollution Impacts

In-Plant	Pollutants
Wastewater from Scrubber	BOD, heavy metals
Boiler Blowdown	TDS
Cooling Water	Heat
(condensers on turbines)	
Offsite	
Runoff from Landfill	BOD, heavy metals
Leachate from Landfill	TDS

thermal pollution resulting from any cooling water used for steam turbine condensers, although this is generally a small effect if the condenser water is added to the effluent of the treatment plant (usually one or two degrees Fahrenheit).

Off-site water pollution impacts from sludge incineration primarily relate to the runoff and the leachate from landfills used for the ash. Because the combustion process usually results in the calcining of lime used in the treatment process, sludge ash is usually highly alkaline, and leachate will exhibit very low heavy metal concentrations. This should be contrasted to the heavy metal solubilization that can occur if sludge ash is landfilled with refuse, or if sludge is landfilled alone prior to complete stabilization. Under these latter conditions, a substantial fraction of heavy metals in the sludge can be dissolved and enter the groundwater.

SUMMARY

In summary, the primary impacts associated with sludge incineration fall in the air pollution and energy categories. Air pollution can be a significant impact from sludge incineration, particularly for older plants equipped with inadequate air pollution control equipment. However, modern plants, or older plants that have been upgraded, produce air emissions that are modest, particularly in view of the relatively small amounts of sludge generated in most treatment plants. For large central plants it may be appropriate to consider designing for air pollution control efficiencies greater than those corresponding to NSPS, particularly in new plants with furnaces operating at low excess air (minimizing the flue gas volume) and those located in areas already severely impacted.

Impacts in the energy area may be minimized or essentially eliminated by the use of available high-performance sludge dewatering devices, such as the belt press and the filter press. Through proper selection of a furnace concept, fossil fuel use can be limited to startup and ash transportation. Energy recovery practiced with sludge incineration can offset in-plant fuel use for heat and hot water and, if appropriate nearby energy markets can be found, may even generate revenue and reduce or eliminate fossil fuel combustion in the boilers of the energy purchaser.

Sludge incineration need not, and should not, be considered as an environmentally unsound approach to sludge management. Proper design and operation to mitigate air pollution and substantially eliminate wasteful energy use is practical and realistic. When coupled with careful landfill design and operation to avoid water pollution from ash disposal, incineration systems provide positive control of health impacts and of impacts on biota associated with sludge disposal.

REFERENCES

1. Niessen, W. R. *Combustion and Incineration Processes - Applications in Environmental Engineering,* (New York: Marcel Dekker, Inc., 1978).
2. Niessen, W. R. In: *Solid Waste Management Handbook,* D. Wilson, Ed. (New York: Van Nostrand-Reinhold Company, 1977), p. 14.
3. Briggs, G. A. "Diffusion Estimation for Small Emissions," ADTL Contribution File No. 79, Air Resources Atmospheric Turbulence and Diffusion Laboratory, NOAA, Oak Ridge, TN (1973).

CHAPTER 17

ULTIMATE DISPOSAL OF HAZARDOUS SLUDGE
VIA SOLIDIFICATION

Robert B. Pojasek, PhD
Roy F. Weston, Inc.
Woburn, Massachusetts

Recent federal regulations issued under the Resource Conservation and Recovery Act (RCRA) [1] no longer exclude from control under subtitle C sewage sludge from publicly owned treatment works (POTW). Pending the future promulgation of a comprehensive sewage sludge regulation, sewage sludge that exhibits any of the specified characteristics of hazardous waste established in the regulations must be managed as a hazardous waste. This new direction taken by the U. S. Environmental Protection Agency (EPA) is bound to change the operation of many of those involved in sludge management.

A representative sample of every sludge must be tested against the following four EPA characteristics: (1) ignitability, (2) corrosivity, (3) reactivity, and (4) extraction procedure (EP) toxicity. If the sludge fails any of these tests it must be declared as hazardous, and the proper EPA notification form must be filed. The sludge will then be assigned an EPA hazardous waste number.

The characteristic that would be failed most often is the EP-toxicity test, especially at a POTW with considerable input from metal finishing and electroplating shops. Another problem is that these inputs may not always be continuous. Thus, on a given day the sludge may be hazardous, while on another day it is nonhazardous. This would create the need for quite a bit of testing unless the generator were to treat the waste as hazardous all the time.

A "hazardous" sludge must be transported, stored, treated and disposed of in a manner prescribed by the EPA hazardous waste rules and regulations. Nonhazardous sludges currently fall under the less restrictive and less expensive subtitle D regulations. Numerous permits will be required for handling hazardous sludges.

Solidification is a treatment process that can operationally convert a "hazardous" sludge into a nonhazardous material, thus avoiding some of the disposal problems. The concept of this conversion is discussed below.

SLUDGE CONVERSION CONCEPT

According to EPA regulations, a hazardous waste ceases to remain hazardous when it does not exhibit any of the characteristics of hazardous waste. If the sludge were to fail the characteristic EP-toxicity, the solidified sludge must be subjected to the same test. The important difference here is the use of the structural integrity procedure (SIP), a test designed to be a moderately severe approximation of the disintegration, which might be expected to occur if a solidified waste were used as structural fill or construction material. Under these conditions, crushing might occur from the passage of heavy equipment over the waste.

Step 1. of the SIP reads as follows:

> "The sample holder should be filled with the material to be tested. If the sample of waste is a large monolithic block, a portion should be cut from the block having the dimensions of a 3.3 cm (1.3 in.) diameter X 7.1 cm (2.8 in.) cylinder. For a fixated waste, samples may be cast in the form of a 3.3 cm (1.3 in.) diameter X 7.1 cm (2.8 in.) cylinder for purposes of conducting this test. In such cases, the waste may be allowed to cure for 30 days prior to further testing."

If the waste is properly set, little breakdown of the solid will take place. The monolith is leached as a solid instead of a finely ground material. This greatly reduced waste surface area coupled with any chemical fixation of the metals to the solid matrix will greatly improve the chances of the material passing the EP-toxicity test. At this point the solidified sludge can be declassified to nonhazardous status.

SOLIDIFICATION TECHNOLOGY

There is wide variety of commercial solidification processes. However, few vendors have extensive experience with sewage sludges. The Waterways Experiment Station (Vicksburg, Massachusetts) has looked at sludge solidification under an interagency agreement with the EPA Solid and Hazardous Waste Research Division in Cincinnati, Ohio. This work is still in progress at this writing.

The most prevalent means of solidifying sludges is with a lime- or Portland cement-based additive. Most probably, a dewatered sludge would be mixed with the additive and the product allowed to harden. Specific details on the application of solidification to wastes in general can be found in the literature [2-7].

Solidification need not be handled by a vendor. Generic means of sludge solidification are well known. However, these procedures and the related equipment need to be adapted to the waste stream, which may involve a considerable effort. On the other hand, vendors can adapt their processes more quickly but may also charge a royalty on the chemical additives used.

Another decision is whether to solidify at the plant site or transport the sludge as a hazardous waste to an offsite processing center. Many considerations go into such a decision. Some vendors offer mobile processing equipment that is designed to make periodic visits to a plant for the purpose of solidifying the collected wastes. However, if the waste is hazardous, a special permit is needed to store it for more than 90 days.

Incineration ashes must be tested to determine whether they are hazardous under the regulations. These materials can also be solidified as the sludges described above.

CONCEPT INSURANCE

Questions are often asked as to whether the solidified waste will hold up over time. To ensure this will happen, physical, leach and accelerated environmental testing may be required on representative samples using established protocols. EPA has sponsored a number of studies to look at some of these questions; unfortunately, most of the solidified wastes were not designed to meet the common point of passing the structural integrity procedure. Varying amounts of additives were used by the participating vendors, making comparisons between processes for a particular waste type very difficult. Furthermore, there is no great amount of concensus in the testing area. However, the American Society of Testing and Materials (ASTM) has recently mounted a special effort to solve this problem.

Accelerated testing is perhaps the most controversial aspect of concept insurance. Most tests that are designed to be representative of environmental conditions take too long to generate useful data. Just as aerospace construction and electrical components are tested for long-term viability with accelerated techniques, a similar approach must be adapted to the wastes. The selection of the proper test should be attempted only after the use or disposal technique is specified.

A suitable landfill design will have to be proposed to handle nonhazardous solidified sludges. Theoretically, a Section 4004 sanitary landfill should be suitable. In no case should a secured landfill (as specified in the regulations) be required. The acceptable alternative will perhaps be a compromise of those two extremes.

There is also a strong possibility that the solidified product could have a recycling application. Some past uses of solidified waste have included the following: land reclamation, road bed aggregate, landfill caps and liners, and artificial reefs. As in the cases above, certain assurance will be required so that the end use will not pose the "hazard" that the raw sludge would pose with respect to EP-toxicity.

CONCLUSIONS

Most of the overviews presented to those involved in sludge management fail to include solidification as an option. The impact of the new regulations with regard to the classification of many sludges as hazardous should change this. Solidification as a treatment process does offer the potential for the sludge generator to declassify the treated waste stream, thus avoiding some of the stringent hazardous waste requirements. Much information is available on the topic so that the sludge manager can make an informed decision on the need to include solidification in the treatment train.

REFERENCES

1. U. S. Environmental Protection. "Hazardous Waste Regulations," *Federal Register* 45(98):33063 (1980).
2. Pojasek, R. B., Ed. *Stabilization/Solidification Processes for Hazardous Waste Disposal* (Ann Arbor, MI: Ann Arbor Science Publishers, Inc., 1978).

3. Pojasek, R. B., Ed. *Stabilization/Solidification Options for Hazardous Waste Disposal* (Ann Arbor, MI: Ann Arbor Science Publishers, Inc., 1978).
4. Pojasek, R. B., Ed. *Impact of Legislation and Implementation of Disposal Management Practices* (Ann Arbor, MI: Ann Arbor Science Publishers, Inc., 1980).
5. Pojasek, R. B., Ed. *New and Promising Ultimate Disposal Options* (Ann Arbor, MI: Ann Arbor Science Publishers, Inc., 1980).
6. Pojasek, R. B. "Solid-Waste Disposal: Solidification," *Chem. Eng.* 87 (August 13, 1979).
7. Pojasek, R. B. "Disposing of Hazardous Chemical Wastes," *Environ. Sci. Technol.* 13:810-814 (1979).

CHAPTER 18

MONITORING THE RESPONSE OF SOILS AND CROPS TO SLUDGE APPLICATIONS

L. E. Sommers and D. W. Nelson

Department of Agronomy
Purdue University
West Lafayette, Indiana

The application of sewage sludge to agricultural land can alter the physical, chemical and microbiological properties of soils. Likewise, the yield and chemical composition of crops grown on sludge-amended soils can be different from those produced using conventional fertilizer materials. Some of these changes in soil or crop properties can be beneficial, whereas others are quite undesirable. For example, application of a specific sludge may elevate the cadmium concentration in crops resulting in a potential threat to human health, while the same sludge may cause a desirable increase in the protein content of a cereal grain. The application of nearly all sludges to soils will result in measurable alterations of soil and crop properties; however, the magnitude of these changes is a function of the composition and rate of sludge applied, initial soil conditions, crop grown and environmental parameters. The objective of this chapter is not to present a comprehensive review of the effects of sludge application on soil and crop properties, but rather to illustrate these effects and to discuss the major criteria for monitoring soil and crops following sludge application.

COMPOSITION OF SEWAGE SLUDGES

The composition of sludge along with the application rate will influence the type and extent of monitoring required following addition to soils. From a land application standpoint, the major constituents in sludge include: (1) pathogens, (2) persistent organics, (3) plant nutrients (N, P, K, etc.), and (4) heavy metals. The concentrations of N, P, K and selected heavy metals in sewage sludges are shown in Table I [1-3]. Numerous other elements are also found in sludges, including other essential plant nutrients (e.g., Ca, Mg, S, Fe, Mo, B, Mn) [1,2,4], nonessential components (e.g., Hg, As, Be) [1,2,4] and numerous organic compounds (e.g., fats, waxes, greases, residual proteins and carbohydrates) [5]. The data presented in Table I indicate that the composition of sludge is quite diverse. A typical sludge contains approximately 3% N, 2% P, 0.3% K, 500 mg Pb/kg, 1600 mg Zn/kg, 800 mg Cu/kg, 100 mg Ni/kg and 15 mg Cd/kg on a dry weight basis. The extreme ranges shown for all elements in Table I amplify the critical need for a sound program for sludge sampling and analysis whenever sludges are being applied to agricultural land. Current approaches for determining the appropriate rate of sludge application to soils are based on the concentrations of organic N, NH_4^+, NO_3^-, P, K, Pb, Zn, Cu, Ni, Cd and PCB in the sludge [6-8].

Table I. Concentration of Selected Constituents in Sewage Sludges[a]

Component	Sommers [1] Range	Sommers [1] Median	Echelberger [2][b] Range	Echelberger [2][b] Median	Chaney [3] Range	Chaney [3] Median
			%			
N	<0.1 – 17.6	3.3	1.0 – 24.7	7.1	---	---
P	<0.1 – 14.3	2.3	---	---	---	---
K	<0.1 – 2.6	0.3	---	---	---	---
			mg/kg			
Pb	13 – 19,700	500	10 – 28,200	451	52 – 4,900	500
Zn	101 – 27,800	1,740	30 – 34,300	1,770	228 – 6,430	1,430
Cu	84 – 10,400	850	178 – 7,700	685	240 – 3,490	790
Ni	2 – 3,520	82	17 – 9,450	141	10 – 1,260	42
Cd	3 – 3,410	16	3 – 1,020	16	1 – 970	13

[a] All data on an oven-dry solids basis.
[b] PCB: range, <0.04 – 241; median, 7.2.

The U. S. Environmental Protection Agency (EPA) has developed regulations concerning the application of sludge on soils used for growing human food-chain crops [9]. Cadmium is the only metal regulated with both annual and cumulative limits set for different soil and crop situations. Limits were also established for the maximum total PCB concentration in sludge (10 mg/kg) where surface applications could be used. All sludges applied to soils growing food-chain crops must undergo a stabilization process (e.g., aerobic or anaerobic digestion, lagoon storage) to reduce the pathogen content in sludges. In addition, public access to sludge-treated areas must be controlled, and root crops consumed raw cannot be grown on treated soils for 18 months after sludge application. The EPA document does not establish monitoring requirements for soils (other than pH) or crops; however, it is based on a management approach to minimizing any adverse impacts of sludge application on cropland. Additional information on the fate of pathogens in soils or on plants can be found in recent reviews [10-12], so will not be covered here.

The major emphasis in monitoring soils and crops has been placed on N, P, K and heavy metals (Pb, Zn, Cu, Ni and Cd). Data on plant-available P and K in soils are needed to determine the amount of supplemental fertilization required for optimal crop yields. One concern from sludge application on soils is the potential for NO_3^- leaching through the profile and possible contamination of groundwaters. Whenever plant-available N in excess of the N required by the crop grown is applied to soils, the NO_3^- not assimilated by the crop could be leached into groundwaters. Widespread NO_3^- leaching may reduce the quality of the groundwater resources in an area and may lead to human and animal health problems (i.e., methemoglobenemia). From a human health standpoint, increased uptake of Cd by crops is a long-term concern, whereas direct ingestion of Pb by livestock grazing on sludge-treated forages could occur. Zn, Cu and Ni are important because of their ability to reduce yields of crops when present in soils at elevated levels. Therefore, this chapter will emphasize monitoring the N and metal concentrations in soils and crops in relation to use of sludges on agricultural land.

MONITORING SOILS TREATED WITH SEWAGE SLUDGE

Soil Sampling

To evaluate the effects of sludge applications on soil properties, it is essential to use a statistically valid approach for soil sampling. Soils are quite heterogeneous, exhibiting both horizontal and vertical variability. Since sludges are commonly applied to the surface of agricultural soils and then incorporated by plowing or disking, soil samples are primarily obtained from the zone of sludge incorporation (i.e., upper 15-25 cm). This is the same depth sampled for routine soil testing to determine fertilizer recommendations. Surface samples can be used for evaluating the effects of sludge application on P, K and metal concentrations in soils because these constituents have been found to be relatively immobile in soils [13,14]. By contrast, samples must be obtained as a function of depth to evaluate the leaching of NO_3^- in soils amended with sludge. The depth of sampling for NO_3^- analysis is site specific, but will typically range from 1 to 5 meters. An alternative approach to evaluate NO_3^- leaching is the use of suction lysimeters or shallow wells (i.e., plastic pipe) for obtaining samples of the soil solution. However, it should be realized that soil solution samples obtained with suction lysimeters may not be totally representative [15].

The natural variability in soil chemical composition plus that caused by either conventional fertilization or sludge application requires the use of a sound soil sampling program. For irrigated soils treated with manure, the coefficient of variation of Cl^- or NO_3^- in the 1.5- to 4.5-meter depth ranged from 16 to 34% when one soil core was obtained from triplicate plots [16]. Subsequent detailed sampling and analysis indicated that 10 soil cores were required for each plot (15 X 15 meters) to provide means within 20% of the true mean. Based on detailed soil sampling of 24 farm fields, the geometric mean was found to be a better representation than the arithmetic mean for NO_3^- in the upper 60 cm [17]. This study concluded that 36 cores should be composited from each field to give results within 20% of the geometric mean. Similarly, analysis of surface samples collected from apparently uniform agricultural soils has shown that the coefficients of variation for plant-available P, K and Mg can vary as follows: P, 23-53%; K, 19-41%; and Mg, 4-28% [18]. The optimum number of samples can be calculated if preliminary data are available for the variability in soil composition. The number of samples can be calculated from [19]:

$$n = t_\alpha^2 S^2 / D^2$$

where   n = number of samples required,
      t = Student's t at the α probability level,
      S = variance based on preliminary samples, and
      D = desired limit (i.e., 1 mg/kg).

The above and additional considerations in soil sampling are discussed by Petersen and Calvin [19].

Metals in Soils

The application of sludge to soils will increase the concentrations of Pb, Zn, Cu, Ni and Cd. All soils contain a natural background concentration of these metals that is a function of the soil parent materials and previous management practices. It is essential that the metal content in soils be determined either prior to sludge application or in an adjacent (i.e., control) untreated area because the background level must be determined and cannot be assumed to fall within published limits [20]. For example, recent research has shown that soils can contain from <1 to 22 mg Cd/kg [21]. Since the maximum allowable Cd addition to soils currently is 20 kg/ha ($\sim$10 mg/kg) [9], the concentration of Cd in soils amended with sludge can be within the range of untreated soils.

Metal analysis of soils involves the determination of total and/or extractable levels, depending on the objectives of the monitoring program. Total analyses utilize either wet digestion or dry ashing to remove organic matter followed by quantitation of the metals in an acidic solution by atomic absorption spectrophotometry [22]. Since total analyses are time-consuming and costly, extractable metals are more commonly determined on a routine basis in soils amended with sludge. In addition, the majority of extractants used for heavy metal analysis were developed to evaluate the availability of essential trace metals (e.g., Zn, Cu) to crops. Therefore, to use these extractants it is necessary to hypothesize that a relatively constant proportion of sludge-borne metals is recovered from a broad range of soils.

Several extractants have been used to evaluate metals in sludge-amended soils. A comparison of some common soil metal extractants is shown in Table II. For Zn, Cu, Ni, Pb and Cd, extractability decreases in the order: $2\ M\ HNO_3 > NH_4OAc$ (pH 4.75) $> NH_4OAc$ (pH 7) $> H_2O$. In general, the solubility of all heavy metals increases with decreasing pH in either the natural soil solution or an extractant [24]. Water is not an appropriate extractant to monitor sludge-amended soils because the amount of metals removed is strongly dependent on the existing soil pH and the soil:solution ratio used, not on the amount of sludge-borne metals added. For example, water-extractable Cd and Pb (Table II) can be similar for both untreated and sludge-treated soils. Similarly, NH_4OAc (pH 7) is used routinely to extract cations associated with the cation exchange complex of soils, while the majority of metals added to soils in sludge are not present as an exchangeable cation [25-26]. The most common metal extractants used for sludge-amended soils are double acid (0.05 N HCl and 0.025 N H_2SO_4), 0.1 N HCl and DTPA.*

---

* DTPA is diethylenetriaminepentaacetic acid. The extractant consists of 0.01 M $CaCl_2$, 0.1 M triethanolamine and 0.005 M DTPA and the solution is adjusted to pH 7.3 [27].

Table II. Extractability of Heavy Metals in Soils Amended with Sewage Sludge [23]

Soil	Extractant[a]	Metal extracted				
		Zn	Cu	Ni	Pb	Cd
		mg/kg				
Untreated	H_2O	0.03	0.24	<0.06	<0.04	<0.01
	NH_4OAc	4.63	0.28	0.13	0.37	0.12
	$NH_4OAc+HOAc$	49.3	0.60	0.28	4.62	0.42
	HNO_3	55	11.2	5.8	27.8	0.69
Sludge-Treated	H_2O	0.09	1.49	0.20	<0.04	<0.01
	NH_4OAc	30.8	3.75	0.44	0.36	0.37
	$NH_4OAc+HOAc$	237.2	8.75	2.43	5.90	1.65
	HNO_3	1084	228	24.4	98.6	2.47

[a] H_2O; H_2O saturated with CO_2; NH_4OAc, 1 M NH_4 acetate at pH 7; NH_4OAc-HOAc, 0.5 M NH_4 acetate + 0.5 M acetic acid at pH 4.75; HNO_3, 2 M HNO_3.

Representative data on the amounts of metals extracted from sludge-amended soils by either 0.1 N HCl or DTPA are shown in Tables III-VI. For research plots, the variability in DTPA-extractable Cd, Zn, Cu and Ni is a function of the amount of metal applied (Table III). For most metals, the coefficient of variation decreases with increasing soil metal concentrations. The greatest variability exists in the nonsludge-treated soils, a result consistent with previous research on the spatial heterogeneity of soil chemical composition. The data in Table III also indicate that increased metal levels in soils can be detected with the DTPA extractant. However, the amounts of Cd added to these experimental plots are considerably in excess of those recommended for agricultural cropland. A greater proportion of Zn and Cd is extracted from sludge-amended soil by 0.1 N HCl than by DTPA (Table IV). The data presented also indicate that both extractants remove increasing amounts of Zn and Cd with increasing sludge application rates. Metal extractability also increased with time after sludge application. A comparison of metals extracted from sludge-amended soils showed that 0.05 N HCl + 0.025 N H_2SO_4 removed a greater proportion of total heavy metals than DTPA [29]. Additional data on the 0.1 N HCl-extractable Zn and Cd concentrations found in soils treated annually with sewage sludge are presented in Table V. Extractable levels of Zn and Cd were found to increase in proportion to the amount of sludge-borne Zn and Cd applied during each of the four years studied. Furthermore, the concentrations of Zn and Cd in corn leaves was directly related to 0.1 N HCl-extractable levels in soil.

Table III. Variability in DTPA-Extractable Metals for Chalmers
Silt Loam Soil (0–15 cm in depth) Amended with Sewage Sludge[a]

Metal	Sludge Applied	Range		Mean	Coefficient of Variation[b]
	metric ton/ha	------- mg/kg -------			%
Cd	0	0.4 -	1.5	0.8	51
	56	1.5 -	6.2	3.7	39
	112	1.8 -	11.4	8.0	36
	224	6.5 -	16.2	12.6	25
	448	8.6 -	30.4	20.2	29
Zn	0	0.3 -	21.2	7.9	78
	56	21.6 -	73.2	49	32
	112	28.8 -	156	105	33
	224	129 -	202	172	13
	448	149 -	256	206	13
Cu	0	3.6 -	16.7	7.9	28
	56	15.5 -	69.4	40.3	39
	112	20.5 -	145.4	92	37
	224	104 -	211	167	15
	448	113 -	310	235	21
Ni	0	1.9 -	6.4	3.6	35
	56	9.5 -	24.3	15.6	26
	112	11.0 -	38	25.5	25
	224	26 -	48	35.9	20
	448	24 -	79	46.3	29

[a] Data from either 14 or 15 replicated field plots.
[b] Coefficient of variation is the standard deviation expressed as a percentage of the mean.

The metal extractant receiving the greatest attention in monitoring soils treated with sludge is DTPA. The DTPA soil test was originally developed for detection of trace metal (Zn, Cu, Mn and Fe) deficiencies in calcareous soils [29] but has been used in many sludge-related studies. Theoretical equilibrium calculations have shown that DTPA along with several other synthetic chelating agents should effectively extract Zn, Cu, Cd, Pb and Ni from soils [31]. The ability of DTPA to extract Zn, Cu and Cd from soils during and after sludge applications is shown in Table VI. The proportion of Zn, Cu and Cd added extracted with DTPA tends to decrease with increasing sludge application rates. In general, 10–50% of the metals applied to soils in sludge are extractable with DTPA; however, the exact percentage extracted depends strongly on the soil and, probably more importantly,

Table IV. Comparison of 0.1 N HCl and DTPA
as Extractants for Sludge-Borne Zn and Cd [28]

Metal Added			Zn Extracted		Cd Extracted	
Zn	Cd	Year	0.1 N HCl	DTPA	0.1 N HCl	DTPA
——— kg/ha/yr ———			————————— mg/kg —————————			
0	0	1974	3.05	1.12	0.14	0.08
		1975	3.12	1.31	0.12	0.08
51	2.9	1974	10.66	4.84	1.20	0.67
		1975	30.87	15.44	2.37	1.51
102	5.8	1974	32.91	14.96	2.52	2.05
		1975	62.94	31.96	5.62	3.18
204	11.5	1974	67.12	32.08	6.91	4.09
		1975	121.78	58.75	11.08	5.98

Table V. Amounts of 0.1 N HCl-Extractable Zn and Cd
in Soils Amended Annually with Sewage Sludge [30]

Metal	Year	Cumulative Applied[a]	Metal Extracted with 0.1 N HCl at Relative Sludge Application Rate of[b]			
			0	0.25	0.5	1.0
		kg/ha	———————— mg/kg————————			
Zn	1971	1806	13	41	98	181
	1972	1905	16	80	150	277
	1973	2122	13	81	118	244
	1974	2358	20	83	157	341
Cd	1971	77	<0.25	1.3	3.6	6.8
	1972	81	0.27	3.1	4.7	12.1
	1973	88	0.29	3.3	6.7	12.9
	1974	101	0.58	3.8	7.3	16.4

[a] Sludge applications were initiated in 1968.
[b] Zn and Cd applied at each treatment level can be calculated by multiplying the cumulative applied and the relative application rate.

on sludge properties. In addition, the predominant metal forms found in sludge-amended soil may change with time to alter the proportion of metals extracted by DTPA [13] (also Tables IV and VI). Several studies have found

Table VI. Effect of Time on DTPA-Extractable Metals
in Soils Amended with Sewage Sludge [26]

| | Sludge application series (metric ton/ha) | | | | | |
| | Single[a] | | | Annual (accumulated)[b] | | |
Year	112	225	450	350	700	1,400
	Zn added, kg/ha					
	149	298	597	379	756	1,518
	% Zn DTPA extractable					
1972	15	16	16	15	16	16
1973	27	38	37	14	27	12
1974	25	25	22	22	25	20
1975	25	17	16	18	13	12
	Cu added, kg/ha					
	26.3	52.6	105	66	132	263
	% Cu DTPA extractable					
1972	40	39	39	40	39	39
1973	61	83	72	29	59	24
1974	56	55	54	55	58	46
1975	53	40	46	48	36	35
	Cd added, kg/ha					
	0.9	1.7	3.4	2.1	4.1	8.3
	% Cd DTPA extractable					
1972	<1	8	4	<1	8	4
1973	4	5	4	20	3	4
1974	7	12	10	8	8	16
1975	1	14	1	14	8	1

[a] Applied in spring 1972.
[b] Applied in three approximately equal applications in spring 1972, fall 1972 and fall 1973.

excellent correlations between DTPA-extractable metals and metal concentrations in plant tissues [13,26,29,32,33]. However, the relationship between DTPA-extractable metals and plant tissue metal concentrations is dependent on the crop grown and soil properties (e.g., pH). Since the DTPA extractant typically removes 25–50% of the metals, this procedure has been used to assess the accumulation of metals in long-term sludge application sites where the rates of application are not known with certainty. For example, DTPA-extractable Cd was found to range from 0.09 to 0.13 mg/kg in nontreated soils and from 0.53 to 7.15 mg/kg in soils used for sludge application by four northeastern U. S. cities [3].

Nearly all data collected indicate that both total and extractable metals are useful parameters to characterize the accumulation of metals in soils amended with sludge. At present, additional information is needed to correlate extractable metals in soils with the resultant concentrations found in the various crops that could be grown on a sludge application site. It is essential that soil samples be collected and analyzed either prior to sludge application or from an adjacent control (nontreated) site to describe the accumulation of sludge-borne metals in soils. For most soils, the relative immobility of sludge-borne metals results in the need to only sample the surface soil (i.e., depth of sludge incorporation).

Nitrogen in Soils

As discussed previously, the main concern from sludge-derived N applications to soils is to minimize the potential for NO_3^- leaching into groundwater. From a management standpoint, minimal amounts of NO_3^- leaching will occur if the amount of plant-available N applied to soil in sludge is equivalent to the N requirements of the crop grown. In liquid sludges containing 2–10% solids, 25–50% of the total N is present as NH_4^+, while the remainder is in organic forms [1]. After application to soils, NO_3^- will be formed via nitrification of the NH_4^+ initially present plus the NH_4^+ formed from mineralization of the organic N. Under optimum temperature and moisture conditions, the majority of the NH_4^+ will be nitrified within one month after sludge application. Nitrogen can also be lost from sludge-amended soils through denitrification (reduction of NO_3^- to N_2 under anaerobic conditions) or NH_3 volatilization (especially from surface applications).

Leaching of NO_3^- has been documented in soils amended with sludges. The application of 1764 kg NH_4^+-N/ha in liquid sludge resulted in NO_3^--N concentrations in soil leachates of >100 mg/l [34]. These excessive NO_3^- levels in the soil solution are to be anticipated because plant-available N added was 10–20 times that required by the crop grown. This study concluded that sludge application rates of $\leqslant 15$ metric ton/ha would minimize the amounts of NO_3^- leached. A study designed to compare NO_3^- levels in soils treated with NH_4NO_3 and digested sludges indicated that similar NO_3^- concentrations were found in soils cropped to bromegrass and treated with 400 kg/ha of nitrogen as NH_4NO_3 and 800 kg/ha of nitrogen as sludge. Since the sludges contained 29–44% of the total N as NH_4^+-N, the amount of plant-available N added to the soils was likely similar for NH_4NO_3 and sludge treatments [35]. As expected, NO_3^- concentrations in 0- to 15-cm soil samples increased with amounts of NH_4NO_3 or sludge N applied. The authors concluded that minimal NO_3^- pollution of groundwater would occur if the amount of N applied in sludge were consistent with the N requirement of the crop grown. Similarly, in soils cropped to corn and treated with a surface application of liquid sludge, soil NO_3^- levels at a 30-cm depth were not increased one year after application for sludge NH_4^+-N additions of 119 kg/ha but were significantly elevated at rates of 237 and 474 kg/ha of NH_4^+-N [36]. The lowest sludge application rate used was recommended because additional amounts of sludge did not increase corn grain yields.

The effect of varying sludge application rates on NO_3^- in soil water at a 120- to 150-cm depth is shown in Table VII. The soils initially were cropped to either rye or sorghum–sudan and then to corn. Minimal increases in soil water NO_3^- concentrations were observed with sludge rates of ≤15 metric ton/ha (i.e., recommended rates for corn). Nitrate-N levels in excess of 100 mg/l were found at the 30 and 60 metric ton/ha rates at several sampling times. However, with lower rates of sludge addition, NO_3^--N concentrations

Table VII. Increases in the Concentration of NO_3^--N in Soil Water
as Affected by Sludge Application Rate and Time [37]

Sludge Applied[a]	Time after application (months)							
	0.75	4	10	12	15	20	22	26
metric ton/ha	————————— mg NO_3^--N/l[b]————————							
	Plano silt loam							
3.75	-7	9	1	4	6	0	-30	--
7.5	1	19	7	0	10	2	-30	--
15	17	17	28	30	-2	44	3	-40
30	40	22	93	80	78	31	0	-20
60	7	35	136	225	101	72	20	0
	Warsaw sandy loam							
3.75	2	13	-14	-29	-14	9	0	-59
7.5	12	18	3	42	9	63	-39	-59
15	18	47	17	12	-9	22	31	-14
30	33	27	33	-14	28	3	0	-3
60	104	125	94	151	110	29	-14	28

[a] Application of 3.75 metric ton/ha added 210 kg/ha organic N, 125 kg/ha NH_4^+-N and 5 kg/ha NO_3^--N. Sludge was surface applied, allowed to dry and incorporated.

[b] Difference between NO_3^--N concentrations of soil water at a depth of 120–150 cm for sludge-amended and control plots.

in soil water were lower than those in control (no N addition) areas. This finding is possible because of the natural variability in soil drainage patterns and in movement of NO_3^- in the soil profile.

The potential impact of sludge additions on groundwater quality can be monitored readily through either NO_3^- analysis of soil cores or soil solution samples. As has been found with inorganic N fertilizers, the potential for NO_3^- leaching into groundwaters always exists when the amount of sludge-borne N applied to soils exceeds the N requirement of the growing crop. Since sludges contain readily available N (NH_4^+) plus mineralizable organic N, both of these sludge N fractions must be considered when selecting an appropriate sludge application rate [7,8]. Because the available data indicate that NO_3^- leaching is comparable for equivalent rates of plant-available N additions in fertilizers or sludges, extensive monitoring of soil profiles for NO_3^- is not needed *provided* that the amount of sludge N applied is consistent with the fertilizer N recommendation for the crop grown.

MONITORING CROPS

The primary concern in monitoring crops grown on soils amended with sewage sludge is the accumulation of heavy metals in plant tissues. Several recent reviews have discussed the behavior of metals in soil–plant systems [38–40]. It is generally concluded that Pb, Zn, Cu, Ni and Cd are the metals of most importance contained in sludges. Cadmium is receiving the greatest attention from the standpoint of its potential long-term impact on human health. Thus, the above five metals will be emphasized here.

Crop Sampling

Several approaches can be used to determine the effect of sludge application on the composition of agronomic crops. To evaluate the impact on animal or human diets, the plant part consumed should be collected and analyzed (i.e., fruit, grain, tuber, forage). A large data base has been established on the chemical composition of diagnostic plant tissues to detect nutrient deficiencies during the growing season. Further, these tissues are being used to monitor the effects of sludge additions on metal content of crops. Table VIII gives for a variety of crops the appropriate diagnostic tissues that should be sampled. Analysis of these tissues is useful in evaluating the effect of sludge application on metal concentrations in crops during their vegetative stage of growth. Data has been compiled on macronutrient and trace element concentrations in a variety of plant species [42]. Even though

Table VIII. Diagnostic Tissues for Field and Vegetable Crops [41]

Crop	Stage of Growth	Plant Part Sampled	N[a]
Corn	Tasseling to silking	Entire leaf at the ear node	15–25
Soybeans and Other Beans	Prior to or during initial flowering	Two or three fully developed leaves at top of plant	40–50
Small Grain	Prior to heading	The four uppermost leaves	50–100
Forage Grasses	Prior to seeding	The four uppermost leaf blades	40–50
Alfalfa	Prior to or at 1/10 bloom	Mature leaf blades from top 2/3 of plant	40–50
Clover and Other Legumes	Prior to bloom	Mature leaf blades from top 2/3 of plant	40–50
Tobacco	Before bloom	Uppermost fully developed leaf	8–12
Sorghum–Milo	Prior to or at heading	Second leaf from top	15–25
Cotton	Prior to or at first bloom	Youngest fully mature leaf on main stem	30–40
Potato	Prior to or during early bloom	Third to sixth leaf from growing tip	20–30
Tomato	Prior to or during early bloom	Third to sixth leaf from growing tip	20–25
Root Crops	Prior to root or bulb enlargement	Center mature leaves	20–30
Leaf Crops	Mid-growth	Youngest mature leaf	35–55
Peas	Prior to or during initial flowering	Leaves from third node down from top of plant	30–60
Sweet Corn	At tasseling	Entire leaf at ear node	20–30

[a] Number of plants to sample.

analysis of diagnostic tissues provides useful information, metal concentrations, particularly Cd, should be determined in the plant part entering the human food chain.

The accumulation of metals in crops is strongly influenced not only by the species grown, but also by the cultivar. The uptake of Cd by different inbred lines of corn and cultivars of soybeans and lettuce is shown in Table IX. The Cd concentration in leaves from various corn hybrids grown on unamended soil can range from <0.06 to 1.49 mg/kg, while sludge addition can result in leaf Cd levels varying from 2.47 to 62.93 mg/kg in different inbreds. Cadmium concentrations in the grain also varied between inbred; however, the values were always less than those found in the leaf for all inbred lines. In a greenhouse experiment, soybean seedlings can contain from 1.4 to 6.0 mg Cd/kg, depending on the cultivar grown. Similarly, lettuce cultivars respond differently to sludge-borne Cd. Analysis of corn grain samples collected from commercial farms throughout Illinois has also revealed

Table IX. Concentrations of Cd in Cultivars of Selected Agronomic Crops

Crop	Inbred Line or Cultivar	Cd in leaf[a]		Cd in grain[a]	
		Ck	SS	Ck	SS
		------ mg/kg ------			
Corn [43][b]	B37	0.92	62.93	0.12	3.87
	H98	1.49	48.84	0.11	2.43
	Mo17	0.59	27.06	0.08	1.20
	R806	0.40	11.38	0.07	0.40
	A632	0.14	11.45	0.11	0.37
	W64A	0.15	9.79	0.06	0.31
	Oh43	0.38	11.33	0.09	0.19
	R805	<0.06	2.47	0.06	0.08
Soybean [44][c]	Richland	--	3.5	--	--
	Jackson	--	3.2	--	--
	Clark '63	--	1.4	--	--
	Grant	--	2.9	--	--
	Mandarin	--	2.0	--	--
	Amsoy '71	--	2.4	--	--
	Corsoy	--	2.2	--	--
	Dunfield	--	4.2	--	--
	Harosoy '63	--	6.0	--	--
	Arksoy	--	4.8	--	--
Lettuce [45][d]	Bibb	1.18	8.40	--	--
	Romaine	0.88	2.25	--	--
	Boston	0.95	3.10	--	--

[a] Ck = unamended soil; SS = sludge amended soil.
[b] Field experiment. Total soil Cd was 0.3 and 21 mg/kg for Ck and SS, respectively.
[c] Greenhouse experiment with 3-week-old plants. Soil contained 7.54 μg/g of 0.1 N HCl-extractable Cd.
[d] Field experiment. Soil contained 0.05 and 1.97 mg/kg of DTPA-extractable Cd.

that background metal concentrations are quite variable in field-grown crops [41]. The concentration of metals in most plant tissues is increased by application of sludge to soil. However, the magnitude is controlled, in part, by the cultivar grown. As shown in Table IX, corn grain Cd was increased by sludge application from 0.12 to 3.87 mg/kg for B37, whereas a minimal change was observed for R805 (0.06 to 0.08 mg/kg).

The above data have important implications from the standpoint of developing a sampling plan to monitor the effect of sludge applications on metal concentrations in crops. As discussed previously for soils, a control area (i.e., crop grown on nontreated soils) must be identified adjacent to the sludge application site. To validly evaluate the effects of sludge application, samples of the crop must be selected concurrently from the control and treated area and be consistent with respect to: (1) cultivar grown, (2) date of collection, (3) plant tissue collected, and (4) cultural practices used in crop production (i.e., fertilization, liming, herbicides). The importance of cultural practices is illustrated by the fact that the Cd content of a crop grown on a non sludge-amended soil at soil pH 5 will likely exceed the Cd concentration grown in a sludge-amended soil at pH 7.5. In addition, the composition of plants will vary from year to year due to the effect of different climatic conditions on plant growth.

Metals in Crops

Plant species differ markedly in uptake of metals from sludge-amended soils. The effects of sewage sludge addition on the concentrations of Zn and Cd in the foliage and edible part of selected vegetables is presented in Table X. For both Zn and Cd, concentrations are typically greater in the foliage than in the edible part. This basic trend has been found to occur for nearly all agronomic crops. Sludge applications (11.2 kg/ha Cd) result in minimal increases in the edible portions of squash, bean, potato and cabbage. In numerous studies, the leafy vegetables (e.g., lettuce, spinach, Swiss chard) have been found to accumulate more Cd than other plants [38]. It should also be noted that crops grown on nonsludge-amended soils contain diverse Cd concentrations in the edible part (0.07–0.54 mg/kg) and foliage (0.16–1.11 mg/kg). Zinc concentrations are similarly increased by sludge additions to soils but may or may not parallel the changes in Cd levels. From the standpoint of minimizing Cd entry into the human diet, crops other than leafy vegetables should be grown on sludge-treated soils.

The Cd, Zn and Cu concentrations in selected agronomic crops are shown in Table XI. As with vegetable crops, metal concentrations are greater in vegetative tissues than in grains. Corn grain is an excellent crop choice for soils treated with sludge because Cd is generally excluded from grain. Both soybeans and oats tend to accumulate more Cd in the grain than corn. For all crops, the vegetative tissue and grain metal concentrations are a function of the amount of metal applied and the prevailing soil conditions (i.e., pH). The concentration of Cu is increased in plants to a smaller extent than either Zn or Cd following sludge application. Even though reduced yields of crops can result from excessive additions of Cu, Zn, Ni or Cd to soils [38,39], the metal application rates shown in Tables X and XI did not cause yield reductions but rather resulted in yield increases. The data presented indicate

Table X. Concentrations of Zn and Cd in Selected
Vegetable Crops Grown on a Sludge-Amended Soil [45]

Crop	Edible part[a]				Foliage[a]			
	Zn		Cd		Zn		Cd	
	Ck	SS	Ck	SS	Ck	SS	Ck	SS
	——————————————— mg/kg———————————————							
Lettuce	54	131	0.30	10.4	––	––	––	––
Sweet Corn	25	40	0.10	1.83	52	142	0.29	16.3
Squash	19	21	0.15	0.27	48	120	0.15	1.40
Pepper	36	45	0.25	1.30	71	96	0.90	7.51
Bean	64	73	0.07	0.21	37	71	0.16	0.72
Broccoli	87	99	0.27	0.89	––	––	––	––
Eggplant	15	22	0.54	1.64	21	22	0.81	2.02
Tomato	26	40	0.52	1.04	38	32	1.11	3.61
Potato	16	19	0.11	0.10	27	33	0.80	0.69
Cabbage	48	59	0.19	0.35	––	––	––	––
Carrot	39	30	0.96	2.29	––	––	––	––
Cantaloupe	18	25	0.21	0.82	––	––	––	––

[a] Ck and SS refer to untreated and sludge-treated soils, respectively. Soil pH ranged from 4.6 to 5.1 and 6.0 to 6.2 for the Ck and SS soils, respectively. Sludge-amended soils received 11.2 kg/ha Cd and 403 kg/ha Zn.

that changes in the metal concentrations of crops can be readily monitored following sludge application on soils. Additional data on crop responses to sludge metal additions can be found in several reviews [38,40,52].

Surface application of sewage sludge on forages can influence the metal levels present through direct contamination in addition to uptake by roots followed by translocation. The application of liquid sludge on a tall fescue pasture results in an immediate increase in the Zn, Cd, Cu, Ni and Pb content because adhering sludge constituted 26–32% of the forage dry weight on the day of application [53] (Table XII). After 44 days, metal concentrations of the forage remained elevated but were considerably less than those found initially. When samples were collected during the following cropping season, no impact of sludge application was observed. From the standpoint of Pb entering livestock feeds, surface application of sludge on forages is likely the most significant route because plant uptake of Pb normally is minimal [38]. Similarly, persistent organics (e.g., PCB) are not absorbed by plant roots and translocated to the shoot, but surface applications of sludge on forages could result in elevated PCB levels in animal feeds resulting in the regulation that sludges containing >10 mg PCB/kg be incorporated into soil [9].

Table XI. Effect of Sludge Application on the
Cd, Zn and Cu Concentrations in Selected Crops

Crop	Metal applied			Metal concentration in						Reference
				Vegetation[a]			Grain			
	Cd	Zn	Cu	Cd	Zn	Cu	Cd	Zn	Cu	
	--- kg/ha ---			------------- mg/kg -------------						
Corn	0	0	--	0.08	19	--	0.005	21	--	28
	11.5	204	--	2.47	76	--	0.180	26	--	
	0	0	0	0.08	19	2.0	0.09	20	0.8	13
	4.3	180	86	0.27	52	3.3	0.10	27	0.7	
	0	0	--	0.3	21	--	0.16	18	--	46
	58.3	1290	--	7.1	112	--	0.44	45	--	
	0	0	--	0.2	59	--	0.09	28	--	30
	101	2358	--	10.9	293	--	0.81	56	--	
	0	0	--	0.42	37	--	<0.05	13	--	47
	64	1523	--	2.04	69	--	<0.05	19	--	
	0	0	0	0.18	31	8.1	0.05	18	1.5	48
	1.8	492	246	0.45	183	12.3	0.12	44	1.8	
	0	0	0	<0.5	24	6	<0.5	39	6	49
	0.3	26	88	0.5	43	7	<0.5	42	3	
Soybeans	0	0	0	0.13	31	5	0.11	45	11	50
	2.4	420	400	0.37	90	5	0.15	69	8	
	0	0	--	0.22	49	--	0.31	49	--	51
	77.6	1632	--	5.78	143	--	0.92	78	--	
	0	0	0	--	145	9	<0.05	76	17	49
	0.3	26	88	--	90	7	<0.05	73	21	
	0	0	--	1.59	42	--	0.41	39	--	47
	64	1523	--	1.80	55	--	0.78	44	--	
Oats	0	0	--	0.77	10	--	0.16	30	--	47
	64	1523	--	1.96	32	--	0.84	39	--	
Sorghum	0	0	--	0.29	23	--	0.03	11	--	28
	11.5	204	--	2.27	47	--	0.36	18	--	
Rye	0	0	0	0.18	22	3.9	--	--	--	13
	4.3	180	86	0.40	56	11.7	--	--	--	
Sorghum–Sudan	0	0	0	0.53	70	6.1	--	--	--	13
	4.3	180	86	0.95	122	9.4	--	--	--	

[a] Vegetation refers to leaf, stover, straw or seedling.

Table XII. Metal Content of Tall Fescue as a Function
of Time after a Surface Application of Sewage Sludge

Time After Sludge Application	Sludge Application Rate	Concentration in tall fescue				
		Zn	Cd	Cu	Ni	Pb
Days	metric ton/ha[a]	$\longleftarrow$ mg/kg $\longrightarrow$				
0	0	11	0.09	4.2	0.24	4.5
	3 (26.2)	244	2.24	95	7.8	88
	6 (31.5)	294	2.72	115	9.4	140
44[b]	3 (5.6)	50	0.47	19	1.6	18
	6 (7.8)	70	0.66	27	2.3	25

[a] Sludge contained 5.9% solids and was applied to unmowed forage; numbers in parentheses are the percentage of tall fescue mass attributable to sludge solids.
[b] Recalculated values.

Table XIII. Effect of Soil pH on Cd Concentrations
in Crops Grown on Soils Amended with Sludge [54]

Site[a]	Treatment	Soil pH	DTPA-Cd[b]	Cd concentration in		
				Swiss Chard	Soybean Grain	Oat Grain
				$\longleftarrow$ mg/kg $\longrightarrow$		
A	Control	5.7	0.13	0.6	0.17	0.05
	Control + lime	6.7	0.14	0.5	0.15	0.04
	Sludge	5.2	0.53	1.9	--	0.23
	Sludge + lime	6.2	0.57	0.6	--	0.07
B	Control	5.3	0.13	3.6	0.36	0.22
	Control + lime	6.7	0.10	1.2	0.28	0.04
	Sludge	4.8	1.13	73.0	3.70	2.12
	Sludge + lime	6.6	1.19	5.5	1.51	0.38
C	Control	5.3	0.93	0.9	0.16	0.11
	Control + lime	6.4	0.96	0.5	0.13	0.07
	Sludge	5.6	7.15	70.4	2.64	3.38
	Sludge + lime	6.6	5.45	17.7	0.65	0.54

[a] Sludge was applied from 1962–1975 at site A, from 1961–1973 at site B, and from 1967–1974 at site C.
[b] Cd extracted from soil by the DTPA extraction procedure [27].

Effect of Soil Properties on Metals in Crops

Several soil properties influence the concentration of metals (primarily Zn, Ni and Cd) in crops, including cation exchange capacity, metal adsorption capacity, organic matter and clay content, and pH. In addition to the total amount of sludge-borne metals applied, pH appears to be the most critical soil property influencing plant uptake of metals [3,38,39,52]. The data presented in Table XIII illustrate the effect of liming acid soils treated with sludge on uptake of Cd by Swiss chard, soybeans and oats. Liming the soils to pH 6.2-6.7 reduced Cd concentrations in all crops, not only in the sludge-treated soils, but also in the controls. Even though liming reduced Cd uptake, crop Cd levels grown in the sludge-treated soils remained above those found in the control treatments. These data also show that the DTPA extraction procedure was useful to detect the increased soil Cd concentrations caused by sludge application. For two of the three soils studied, liming did not affect the amounts of Cd extractable by DTPA. It is apparent that the Cd concentrations in plants are influenced by both the total amount of Cd applied to the soil and soil pH. In general, increases in Cd concentrations of crops will be minimized by maintaining soil pH at 6.5 or above [38], as required by the EPA whenever food-chain crops are grown [9].

SUMMARY

The effect of sewage sludge applications on the chemical composition of soils and crops can be readily monitored. Because of the natural horizontal and vertical heterogeneity of soils, a sound soil sampling program is needed to obtain statistically valid data on sludge-induced changes in soil properties. Similarly, the concentrations of many macro- and micronutrients and heavy metals in crops are increased by sludge application to soils. Since crop species and cultivar of the same species differ markedly in metal uptake, samples of crops must be obtained from both sludge-treated and control areas to evaluate the impact of sludge applications on plants. Emphasis has been placed on analyzing metals in the edible part of plants (e.g., grain, fruit, tuber or forage) to obtain data relative to human health. The extent of soil and crop monitoring required depends on the sludge application rates used. If sludge application rates are consistent with the amount of N required by the crop [7,8] and the limitations on Cd and sludge stabilization [9], no monitoring of crops is needed. Since soil pH is the primary factor governing plant uptake of heavy metals (e.g., Zn, Cu, Ni and Cd), soils must be monitored routinely to ensure that the pH is 6.5 or above whenever food-chain crops are grown [9]. In addition, routine soil testing for available P and K is recommended so that supplemental fertilizer P and K can be applied, if needed, to optimize crop yields. More extensive soil and crop monitoring is required if sludge application rates exceed current recommendations [7-9] for agricultural land, the major considerations being Cd accumulation in crops; NO_3^- leaching into groundwaters; accumulation of Pb, Zn, Cu and Ni in soils; and maintenance of soil pH at 6.5 or above.

ACKNOWLEDGMENTS

This research was supported, in part, by Western Regional project (W-124), entitled "Optimum Utilization of Sewage Sludge on Agricultural Land" and published with the approval of the Director of the Indiana Agricultural Experiment Station, Purdue University, as Journal Paper No. 8109.

REFERENCES

1. Sommers, L. E. "Chemical Composition of Sewage Sludges and Analysis of Their Potential Use as Fertilizers," *J. Environ. Qual.* 6:225-232 (1977).
2. Echelberger, W. F., Jr., J. M. Jeter, F. P. Girardi, P. M. Ramey, G. Glenn, D. Skole, E. Rogers, J. C. Randolph and J. Zogorski. "Municipal and Industrial Wastewater Sludge Inventory in Indiana; Chemical Characterization of Municipal Wastewater Sludge in Indiana, Part 1," School of Public and Environmental Affairs, Indiana University, Bloomington, IN (1979).
3. Chaney, R. L., S. B. Hornick and P. W. Simon. "Heavy Metal Relationships During Land Utilization of Sewage Sludge in the Northeast," in *Land as a Waste Management Alternative,* R. C. Loehr, Ed. (Ann Arbor, MI: Ann Arbor Science Publishers, Inc., 1977), pp. 283-314.
4. Furr, A. K., A. W. Lawrence, S. S. C. Tong, M. C. Grandolfa, R. A. Hofsteader, C. A. Bache, W. H. Gutemann and D. J. Lisk. "Multielement and Chlorinated Hydrocarbon Analysis of Municipal Sewage Sludges of American Cities," *Environ. Sci. Technol.* 10:683-687 (1976).
5. Strachan, S. D. "Characterization of Organic Components in Sewage Sludge," Unpublished M. S. Thesis, Purdue University, West Lafayette, IN (1978).
6. U. S. Environmental Protection Agency. "Municipal Sludge Management: Environmental Factors," MCD-28, EPA 430/9-77-004, Washington, DC (1977).
7. Knezek, B. D., and R. H. Miller. "Application of Sludges and Wastewaters on Agricultural Land: A Planning and Educational Guide," North Central Regional Publication 235, Ohio Agricultural Research and Development Center, Wooster, OH (1976). Reprinted by U. S. Environmental Protection Agency as MCD-35 (1978).

8. "Principles and Design Criteria for Sewage Sludge Application on Land," in *Sludge Treatment and Disposal, Part 2, Sludge Disposal* (Cincinnati, OH: Environmental Research Information Center, U. S. Environmental Protection Agency, 1978), pp. 57-112.

9. "Criteria for Classification of Solid Waste Disposal Facilities and Practices; Final, Interim Final, and Proposed Regulations" (as corrected in the *Federal Register* of September 21, 1979), *Federal Register* 44: 53438-53469 (1979).

10. Sagik, B. P., and C. E. Sorber. "Risk Assessment and Health Effects of Land Application of Municipal Wastewater and Sludges," Center for Applied Research and Technology, University of Texas at San Antonio, San Antonio, TX (1978).

11. Burge, W. D., and P. B. Marsh. "Infectious Disease Hazards of Land Spreading Sewage Wastes," *J. Environ. Qual.* 7:1-9 (1978).

12. Elliott, L. F., and J. R. Ellis. "Bacterial and Viral Pathogens Associated with Land Application of Organic Wastes," *J. Environ. Qual.* 6:245-251 (1977).

13. Kelling, K. A., D. R. Keeney, L. M. Walsh and J. A. Ryan. "A Field Study of the Agricultural Use of Sewage Sludge: III. Effect on Uptake and Extractability of Sludge-Borne Metals," *J. Environ. Qual.* 6: 353-358 (1977).

14. Sommers, L. E., D. W. Nelson and D. J. Silviera. "Transformations of Carbon, Nitrogen, and Metals in Soils Treated with Waste Materials," *J. Environ. Qual.* 8:287-294 (1979).

15. Hansen, E. A., and A. R. Harris. "Validity of Soil-Water Samples Collected with Porous Ceramic Cups," *Soil Sci. Soc. Am. Proc.* 39:528-536 (1975).

16. Pratt, P. F., J. E. Warneke and P. A. Nash. "Sampling the Unsaturated Zone in Irrigated Field Plots," *Soil Sci. Soc. Am. J.* 40:277-279 (1976).

17. Reuss, J. O., P. M. Soltanpour and A. E. Ludwick. "Sampling Distribution of Nitrates in Irrigated Fields," *Agron. J.* 69:588-592 (1977).

18. Leo, M. W. M. "Heterogeneity of Soil of Agricultural Land in Relation to Soil Sampling," *J. Agric. Food Chem.* 11:432-434 (1963).

19. Peterson, R. G., and L. D. Calvin. "Sampling," in *Methods of Soil Analysis, Part II.* C. A. Black, Ed. (Madison, WI: American Society of Agronomy, 1965), pp. 54-72.

20. Allaway, W. H. "Agronomic Controls Over the Environmental Cycling of Trace Elements," *Advan. Agron.* 20:235-274 (1968).

21. A. L. Page, Department of Soil and Environmental Sciences, University of California, Riverside. Personal communication (1980).

22. Ellis, R., Jr., J. J. Hanway, G. Holmgren, D. R. Keeney and O. W. Bidwell. "Sampling and Analysis of Soil, Plants, Wastewaters, and Sludge: Suggested Standardization and Methodology," North Central Regional Publication 230, Kansas Agricultural Experiment Station, Manhattan, KS (1975).

23. Wiklander, L., and K. Vahtras. "Solubility and Uptake of Heavy Metals From a Swedish Soil," *Geoderma* 19:123-129 (1979).

24. Lindsay, W. L. *Chemical Equilibria in Soils.* (New York: John Wiley & Sons, Inc., 1979).

25. Silviera, D. J., and L. E. Sommers. "Extractability of Copper, Zinc,

Cadmium, and Lead in Soils Incubated with Sewage Sludge," *J. Environ. Qual.* 6:47-52 (1977).

26. Latterell, J. J., R. H. Dowdy and W. E. Larsen. "Correlation of Extractable Metals and Metal Uptake of Snapbeans Grown on Soil Amended with Sewage Sludge," *J. Environ. Qual.* 7:435-440 (1978).

27. Lindsay, W. L., and W. A. Norvell. "Development of a DTPA Micronutrient Soil Test for Zinc, Iron, Manganese, and Copper," *Soil Sci. Soc. Am. J.* 42:421-428 (1978).

28. Baker, D. E., M. C. Amacher and W. T. Doty. "Monitoring Sewage Sludges, Soils, and Crops for Zinc and Cadmium," in *Land as a Waste Management Alternative*, R. C. Loehr, Ed. (Ann Arbor, MI: Ann Arbor Science Publishers, Inc., 1977), pp. 261-281.

29. Korcak, R. F., and D. S. Fanning. "Extractability of Cadmium, Copper, Nickel, and Zinc by Double Acid Versus DTPA and Plant Content at Excessive Soil Levels," *J. Environ. Qual.* 7:506-512 (1978).

30. Hinesly, T. D., R. L. Jones, E. L. Ziegler and J. J. Tyler. "Effects of Annual and Accumulative Applications of Sewage Sludge on the Assimilation of Zinc and Cadmium by Corn (*Zea mays* L.)," *Environ. Sci. Technol.* 11:182-188 (1977).

31. Sommers, L. E., and W. L. Lindsay. "Effect of pH and Redox on Predicted Heavy Metal-Chelate Equilibria in Soils," *Soil Sci. Soc. Am. J.* 43:39-47 (1979).

32. Street, J. J., W. L. Lindsay and B. R. Sabey. "Solubility and Plant Uptake of Cadmium in Soils Amended with Cadmium and Sewage Sludge," *J. Environ. Qual.* 6:72-77 (1977).

33. Mitchell, G. A., F. T. Bingham and A. L. Page. "Yield and Metal Composition of Lettuce and Wheat Grown on Soils Amended with Sewage Sludge Enriched with Cadmium, Copper, Nickel, and Zinc," *J. Environ. Qual.* 7:165-171 (1978).

34. Hinsley, T. D., O. C. Braids and J. A. E. Molina. "Agricultural Benefits and Environmental Changes Resulting From the Use of Digested Sewage Sludge on Field Crops," an interim report on a solid waste demonstration project. U. S. Environmental Protection Agency, Washington, DC (1971).

35. Soon, Y. K., T. E. Bates, E. G. Beauchamp and J. R. Moyer. "Land Application of Chemically Treated Sewage Sludge: I. Effects on Crop Yield and Nitrogen Availability," *J. Environ. Qual.* 7:264-269 (1978).

36. Stewart, N. E., E. G. Beauchamp, C. T. Corke and L. R. Webber. "Nitrate Nitrogen Distribution in Corn Land Following Applications of Digested Sewage Sludge," *Can. J. Soil Sci.* 55:287-294 (1975).

37. Kelling, K. A. L. M. Walsh, D. R. Keeney, J. A. Ryan and A. E. Peterson. "A Field Study of the Agricultural Use of Sewage Sludge: II. Effect on Soil N and P," *J. Environ. Qual.* 6:345-352 (1977).

38. "Application of Sewage Sludge to Cropland: Appraisal of Potential Hazards of the Heavy Metals to Plants and Animals," Council for Agricultural Science and Technology Rept. No. 64, Council for Agricultural Science and Technology, Ames, IA (1976). Reprinted by U. S. Environmental Protection Agency as MCD-33 (1976).

39. Chaney, R. L., and P. M. Giordano. "Microelements as Related to Plant Deficiencies and Toxicity," in *Soils for Management of Organic*

Wastes and Wastewaters, L. F. Elliott and F. J. Stevenson, Eds. (Madison, WI: Soil Science Society of America, 1977), pp. 234-279.

40. Baker, D. E., and L. Chesnin. "Chemical Monitoring of Soils for Environmental Quality and Animal and Human Health," *Advan. Agron.* 27:305-374 (1976).

41. Walsh, L. M., and J. D. Beaton, Eds. *Soil Testing and Plant Analysis* (Madison, WI: Soil Science Society of America, 1973).

42. Pietz, R. I., J. R. Peterson, C. Lue-Hing and L. F. Welch. "Variability in the Concentration of 12 Elements in Corn Grain," *J. Environ. Qual.* 7:106-110 (1978).

43. Hinesly, T. D., D. E. Alexander, E. L. Ziegler and G. L. Barrett. "Zinc and Cadmium Accumulation by Corn Inbreds Grown on Sludge Amended Soil," *Agron. J.* 70:425-428 (1978).

44. Boggess, S. F., S. Willavize and D. E. Koeppe. "Differential Response of Soybean Varieties to Soil Cadmium," *Agron. J.* 70:756-760 (1978).

45. Giordano, P. M., D. A. Mays and A. D. Behel, Jr., "Soil Temperature Effects on Uptake of Cadmium and Zinc by Vegetables Grown on Sludge-Amended Soil," *J. Environ. Qual.* 8:233-236 (1979).

46. Hinesly, T. D., E. L. Ziegler and G. L. Barrett. "Residual Effects of Irrigating Corn with Digested Sewage Sludge," *J. Environ. Qual.* 8: 35-38 (1979).

47. Sommers, L. E., and D. W. Nelson, Department of Agronomy, Purdue University, West Lafayette, IN. Unpublished results.

48. Shaeffer, C. C., A. M. Decker, R. L. Chaney and L. W. Douglass. "Soil Temperature and Sewage Sludge Effects on Metals in Crop Tissue and Soils," *J. Environ. Qual.* 8:455-459 (1979).

49. Ritter, W. F., and R. P. Eastburn. "The Uptake of Heavy Metals from Sewage Sludge Applied to Land by Corn and Soybeans," *Commun. Soil Sci. Plant Anal.* 9:799-811 (1978).

50. Ham, G. E., and R. H. Dowdy. "Soybean Growth and Composition as Influenced by Soil Amendments of Sewage Sludge and Heavy Metals: Field Studies," *Agron. J.* 70:326-330 (1978).

51. Hinesly, T. D., R. L. Jones, J. J. Tyler and E. L. Ziegler. "Soybean Yield Responses and Assimilation of Zn and Cd from Sewage Sludge-Amended Soil," *J. Water Poll. Control Fed.* 48:2137-2152 (1976).

52. Kirkham, M. B. "Trace Elements in Sludge on Land: Effect on Plants, Soils, and Ground Water," in *Land as a Waste Management Alternative,* R. C. Loehr, Ed. (Ann Arbor, MI: Ann Arbor Science Publishers, Inc., 1977), pp. 209-247.

53. Chaney, R. L., and C. A. Lloyd. "Adherence of Spray-Applied Liquid Digested Sewage Sludge to Tall Fescue," *J. Environ. Qual.* 8:407-411 (1979).

54. Chaney, R. L., and S. B. Hornick. "Accumulation and Effects of Cadmium on Crops," in *Proc. First Int. Cadmium Conf.* (London: Metals Bulletin, Limited, 1978), pp. 125-140.

CHAPTER 19

A DISCUSSION OF
GROUNDWATER MONITORING

Lawrence N. Halfen, PhD, General Manager
Environmental Data, Inc.
Grand Rapids, Michigan

Effective groundwater monitoring is a multifaceted enterprise involving attention to technical detail in every phase. Each of the phases—well site selection, casing and screen installation, water sampling, analytical procedures and data interpretation—must be done in a fashion appropriate to achieve the desired goal.

Entire volumes have been devoted to each of the above activities. This chapter will treat selected experiences and observations derived from the groundwater monitoring experiences of a private professional consulting firm engaged in studies related to groundwater and how various sludges impact this resource.

Groundwater is not a static commodity; rather, it is part of a hydrologic cycle involving recharge and discharge of water stored in subsurface horizons. This concept should be kept in the forefront of a groundwater study to avoid the pitfalls of incomplete data and inaccurate interpretations, even when great care is taken in other aspects of the program.

The most important application of this axiom occurs early in all groundwater monitoring programs, when the locations for wells are selected. Factors that influence the hydrologic cycle include, but are not limited to, total and periodic precipitation, runoff, soil permeability, evapotranspiration, subsurface strata and temperature. Other variables frequently apply also, including the nature and amount of contaminants, if any are present; local land use and development; and access to the site. Clients react positively when their groundwater consultant demonstrates his expertise by obtaining relevant and valid data from a minimum number of wells. Not only can the operational alternatives be considered promptly, but there is the added benefit of cost-effectiveness.

ROLE OF REGULATORY AGENCIES

The groundwaters of the state of Michigan are protected by P. A. 245 [1] of 1929, commonly known as the Michigan Water Resources Commission Act. This law and the associated rules and regulations package proposed by the Michigan Department of Natural Resources (DNR) clearly state that any deterioration of surface water or groundwater quality is not acceptable and subject to legal remedy.

Sludge management, which includes direct application of sludge or sludge by-products to the land, may influence groundwater by adding dissolved or suspended materials. This is due, in part, to the ability of the polar water molecule to solubilize a remarkable variety of atoms, ions and molecules, thereby assuring the movement of some sludge components to various saturated strata. The carrying capacity in this solvent system is such that the passive dispersion of nonpolar (hydrophobic) species of molecules to the water table occurs with only the minimum necessary solvation.

The rate of movement of specific sludge components will be affected by the composition, absorption characteristics and number of soil horizons these materials must infiltrate to reach a static water level. This increased concentration of undesirable ions, molecules and/or microorganisms may result in a deterioration of groundwater quality.

In instances in which groundwater is used as the source of drinking water, negative impacts on groundwater quality are the concern of the Michigan Department of Public Health. P. A. 399 [2] (1976), also known as the Safe Drinking Water Act, gives this agency the authority to set standards for groundwater quality in drinking water applications.

P. A. 245 and 399 form the basis of groundwater concerns by the government of the state of Michigan. To be effective in protecting and managing this resource, public regulation must be based on adequate programs of analytical monitoring using realistic standards directly linked to modern research findings. Both state and federal agencies have expended considerable efforts to provide coherence and definition for groundwater testing to generate a meaningful data base for groundwater management decisions.

The U. S. Environmental Protection Agency (EPA), under the mandate of the Safe Drinking Water Act [3] of 1974 establishes the National Primary Interim Drinking Water Regulations [4], which set drinking water (groundwater) evaluation standards for the evaluation of drinking water. These include inorganic, organic, bacteriological, turbidimetric, and radiological parameters. Figure 1 presents this list of parameters. The levels associated with them are regarded as a "maximum level" on a total constituent basis. This list was quickly adopted by many states, including Michigan, as a basis for a drinking water statute and as an action guideline for agency representatives responsible for the environmental and health regulations. Figure 2 (a-c) lists those parameters currently used by the Michigan Department of Natural Resources [5] in its management of drinking water (groundwater). Most of these informal standards are borrowed from other sources, including the National Primary Interim Drinking Water Regulations.

CONTAMINANT   MAXIMUM LEVEL

Inorganic Chemicals (Total constituent values)

Arsenic	0.05	mg/l
Barium	1	mg/l
Cadmium	0.010	mg/l
Chromium	0.05	mg/l
Lead	0.05	mg/l
Mercury	0.002	mg/l
Nitrate (as N)	10	mg/l
Selenium	0.01	mg/l
Silver	0.05	mg/l
Fluoride	1.4-2.4	mg/l

Organic Chemicals

Endrin	0.0002	mg/l
Lindane	0.004	mg/l
Methoxychlor	0.1	mg/l
Toxaphene	0.005	mg/l
2,4-D	0.1	mg/l
2,4,5-TP Silvex	0.01	mg/l
Turbidity	1 TU-5 TU	
Coliform Bacteria	<1/100 ml (mean)	

Radiological

Radium -226 and -228	5pCi/l
Gross Beta	4 mrem/year (50 pCi/l)
Gross Alpha	15 pCi/l

Further details are found in the Federal Register.

Figure 1. Current contamination limits and test procedures set by interim primary regulations (SDWA) for drinking water.

The wide adoption of the federal standards is understandable considering the difficulty in determining which substances should be classified as water contaminants and at what levels they become important. Many substances in Figure 1 appear there because exposure data were available when the rules were established. More attention should be given to the development of a comprehensive approach to these standards. For example, a series of decreasingly important pesticides are on the list, yet demonstrated carcinogens such as vinylchloride and benzene are not. In addition, research results suggest the need for review and new definition in the more traditional public health approach to drinking water bacteriology [6].

New problems regulating groundwater result when improvements in analytical methodologies and instrumentation push detection limits to even lower levels. Thus, because of diminishingly small levels of a contaminant, research results related to the impact on living organisms are much slower in appearing. The regulatory agency is then confronted with a dilemma in deciding whether laboratory findings of a ppm, ppb or ppt of a contaminant are significant. Analytical methods provide the numbers, but not the interpretation of those values.

Parameter	Concentration Level	Application	Source
Alkyl Benzene Sulfonate (ABS)	0.5 mg/l	Mandatory	USPHS
Alkalinity	—Not a specific-polluting substance		
Ammonia (NH$_4$)	0.10 mg/l	Recommended	USPHS
Arsenic	0.01 mg/l	Recommended	USPHS
	0.05 mg/l	Mandatory	USEPA
Barium	1.0 mg/l	Mandatory	USPHS USEPA
Bicarbonate	150 mg/l	Recommended	Hibbard
Cadmium	0.01 mg/l	Mandatory	USPHS USEPA
Calcium	75 mg/l	Recommended	WHO
Carbon Chloroform Extract	0.2 mg/l	Recommended	USPHS
Carbonate	20 mg/l	Recommended	Hibbard
Chloride	250 mg/l	Recommended	USPHS
Chlorine	—Very high levels will be objectionable		USEPA
Chromium (Hexavalent)	0.05 mg/l	Mandatory	USPHS USEPA
COD	<50 ppm	Recommended	USEPA
Color	15 Units	Recommended	USPHS
Conductivity	—Function of dissolved solids.		
Copper	1.0 mg/l	Recommended	USPHS USEPA
Corrosivity	—Non-Corrosive		USEPA
Cyanide	0.01 mg/l	Recommended	USPHS
	0.2 mg/l	Mandatory	USEPA

Figure 2. (a) MDNR working guidelines for drinking water quality.

Parameter	Concentration Level	Application	Source
Dissolved Solids	500 mg/l	Recommended	USPHS
Dissolved Oxygen	No Limits	n/a	USPHS
Foaming Agents	0.5 mg/l	Recommended	USEPA
Hardness as CaCO$_3$	300 ppm	Recommended	USPHS USEPA
Herbicides	Not Established	n/a	MDNR
Hydrocarbons	0.2 ppm	Recommended	MDNR
Hydrogen Sulfide	0.05	Recommended	USEPA
Iron	0.3 mg/l	Recommended	USPHS USEPA
Lead	0.05 mg/l	Mandatory	USPHS USEPA
Magnesium	No Limits	n/a	USPHS
Magnesium & Sodium Sulfate	500 mg/l 1000 mg/l	Permissible Excessive	WHO WHO
Manganese	0.05 mg/l	Recommended	USPHS USEPA
Mercury	0.002 mg/l	Mandatory	USEPA
Nickel	0.1 mg/l	Recommended	USEPA
Nitrate (as N)	10.0 mg/l	Mandatory	USEPA
Nitrate (as NO$_3$)	45 mg/l	Recommended	USPHS
Nitrogen Ammonia (as NH$_4^+$)	0.5 mg/l	Recommended	USPHS
Nitrogen Nitrite	2 mg/l	Recommended	MDNR

Figure 2. (b) MDNR working guidelines for drinking water quality.

Parameter	Concentration Level	Application	Source
Nitrogen Organic	No Limits	n/a	MDNR
Odor	#3	Mandatory	USPHS
PCB	No Limits	n/a	MDNR
Pesticides	Not Established	n/a	MDNR
Phenolic Compounds (as Phenols)			
pH	6.5-8.5	Recommended	USEPA
Phosphate	0.001 mg/l	Mandatory	USPHS USEPA
Potassium	1000-2000 mg/l	Extreme Limit	MDNR
Selenium	0.01 mg/l	Mandatory	USPHS USEPA
Silica	No Limits	n/a	MDNR
Silver	0.05 mg/l	Mandatory	USPHS USEPA
Sodium	10 mg/l	Recommended	Hibbard
Sulfate	250 mg/l	Recommended	USPHS USEPA
Tannates (See Color)			
Total Dissolved Solids	500 mg/l	Recommended	USPHS USEPA
TOC	<50 mg/l	Recommended	MDNR
Turbidity	5 Units	Mandatory	USPHS
Zinc	5 mg/l	Recommended	USPHS USEPA

Figure 2. (c) MDNR working guidelines for drinking water quality.

These circumstances place the regulatory agency in a difficult position. There is a clear mandate to protect the environment for the public good. The laws and regulations assure that these resources are not to be used for the disposal or treatment works of private interests at public expense and/or peril. Agencies note people's concern about environmentally induced cancer and carcinogenic waste in groundwater.

While due consideration must be given to detection levels for hazardous and/or carcinogenic substances, regulatory agency decisions must be based on substantive identification of negative effect levels. To do otherwise poses technical and financial problems, especially in large or older contamination incidents involving substantial volumes of water. Few data address detoxification of compounds in biological systems and the establishment of "threshold" values that would preclude the negative effects of a substance. Chemical trade associations are now expressing concern over the extent of environmental hazard posed by certain detectable compounds in the environment [7,8].

The position of an environmental consultant is equally dificult under these circumstances. Certainly an environmental professional would not advocate the release of cancer-causing agents or microbial disease vectors; however, the consultant also has a responsibility to provide his client and the public with an effective, timely and practical solution to groundwater contamination problems at a reasonable cost. This goal is frequently frustrated when treatment and discharge limits are regarded as acceptable only when they are nondetectable, rather than below the threshold level of effect of the compound in actual living systems. Effort should now be focused on the validity of testing and research methods that originally classified many chemical species as hazardous or carcinogenic.

This overview serves to show that the environmental consultant is positioned between those who must protect a resource within the framework of laws and rules and private interests who must function in a profitable and cost-effective fashion. It is in this context that groundwater monitoring is conceived and conducted.

GROUNDWATER MONITORING

Effective groundwater monitoring requires the coordination of four areas: sampling, parameter selection, accurate analysis and competent interpretation of the results. Aspects of these four areas are discussed below.

Sampling

There are several ways of sampling groundwater to determine its characteristics, all of which have some disadvantages. Bailers and thief samplers (Figure 3) are simple and quick. Sometimes, however, they collect condensed contamination on the inside of the well casing when they are withdrawn. Also, in wells where the screen is quite deep there is little chance to obtain a representative sample of water as it enters the casing from the water table. While this method of sampling is particularly valuable on small monitor wells, it must be preceded by sufficient bailing of the well casing to ensure renewal from outside the well.

Some firms use the airlift method for well development, which can quickly result in water production from even very deep wells with spectacular results (Figure 4). This procedure involves the rapid introduction of large volumes of air into the casing and may result in the volatilization of low-molecular-weight organic contaminants. Samples from this method will be atypical

Figure 3. Hand sampling a monitor well with a bailer.

Figure 4. An airlift system in use to develop a monitor well.

of the groundwater and frequently give incorrect low values. This method can produce acceptable results provided the values sought are not stripped by vigorous action.

Shallow wells can be easily sampled using a peristaltic pump (Figure 5) if the head is not over 20 feet. This sytem has a low delivery volume but does produce acceptable samples where the well screen is close to the top of the water table. In shallow water tables with a substantial column of water in the casing above the screen, particular care must be exercised to draw off the water in the casing prior to collecting the sample. This ensures that the sample is representative of the water table. The peristaltic pump is readily portable and easy to operate, which is particularly important under adverse field conditions.

Figure 5. A peristaltic pump used to sample a shallow monitor well.

As its name implies, deep-well jet pumps (Figure 6) are used in situations where a peristaltic pump is unable to pull a column of water up to the sampling level. The major difficulty associated with the jet pump is that outside water must be added to initiate operation. This addition may alter the overall composition of the wellwater. To compensate for this it is advisable to operate the pumping system until several equivalents of the original priming volume of water have been expelled from the casing, thus ensuring a minimal margin for error. Obviously, the water and equipment introduced into the well must be free of contamination so the sampling activities themselves do not create problems. In comparison with other sampling methods the deep-well jet pump can provide consistently acceptable samples.

Other problems arise from materials used as well casings and sample containers. Three kinds of casing are commonly used: black iron, galvanized

Figure 6. A jet pump used for sample collection from a deep monitor well.

steel and polyvinyl chloride (PVC). Caution should be exercised in using iron or galvanized pipe where acid materials are involved or could be generated. This also applies in those situations in which parameters involved include iron or zinc. Similarly, sample anomalies have been noted when organic determinations were conducted on waters from PVC pipes. Well screens are usually of brass construction with soldered joints; however, increasing numbers of stainless steel well screens are being specified.

Appropriate cleaning and pretreatment of casings to remove oil coatings, soldering fluxes, joint compound and lubricants is a critical detail in obtaining good quality water samples. This may necessitate thorough cleaning of drilling equipment fron one hole to the next, particularly where contaminated subsurface strata are encountered.

Parameter Selection

The selection of laboratory tests to be performed on groundwater samples is important because the quality of the groundwater and how it is impacted are based on the sample data. As discussed earlier, the number of parameter listings at state and federal levels is increasing. The groundwater quality rules proposed for P. A. 245 [9] of 1929 are a case in point. Specific analyses encompassing general water quality are required, including specific conductance, calcium, sodium, magnesium, chloride, sulfate and bicarbonate. Depending on the association of the sample point to industry, municipal treatment plants or landfill disposal activities, additional procedures are also required on general groupings such as organic compounds, toxic materials, metals and hazardous materials. The interpretation of these generalities can range from a single evaluation, such as chemical oxygen demand or total organic carbon, up to the entire contents of the *Michigan Critical Materials Register* [10]. The final list of parameters, which forms the basis of a groundwater monitoring program, usually is the proposed rules for Act 245, information supplied by the client, the specific regulatory agency concerns, and the experience and expertise of the hydrogeological consultant.

The nature and number of parameters to be evaluated has an important influence on sampling methods and subsequent sample handling. This includes the number of samples collected, the size of a sample, the container used, allowable delays prior to analysis, methods of preservation, and associated detail of labeling and packaging. It is only after these procedures have been proposed, agreed to and implemented that a sample will arrive in a laboratory for analysis.

Laboratory Analysis

Laboratory analysis results in bottles of water being transformed into tables of numbers associated with chemical notations on the laboratory data work sheet. First, the sample's arrival is recorded (Figure 7) and all documentation and directions associated with the analytical program are reviewed. Such accurate, detailed and timely records are essential to the operation of any modern laboratory facility. A sample code number is assigned and is keyed to all internal records. Following the analysis, records retain their value as a permanent resource for technical, legal and/or historical interests.

Modern laboratories in which groundwater analyses are conducted are similar with respect to instrumentation and methodologies, since procedures for analysis of various parameters have been standardized and published

Figure 7. Logging a sample into the laboratory with assignment of a code number.

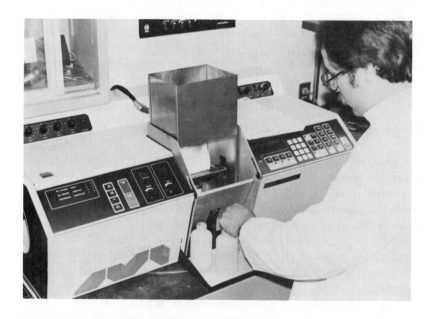

Figure 8. An atomic absorption spectrophotometer in operation.

by such sources as *Standard Methods* [11], EPA [12] and ASTM [13]. New procedures and instrumentation are rapidly appearing, sometimes rendering state-of-the-art practices obsolete before the warranty expires on the instrument involved. Detection limits and sensitivities have improved to the extent that previously acceptable laboratory water supplies are now inadequate for reagent preparation related to modern analysis methods. Atomic absorption spectrophotometers (Figure 8) now routinely quantify metals in the parts per billion range using graphite rod furnaces and microprocessors to provide appropriate background corrections and statistical compensation. This is combined with increasing versatility in sample handling and reduced turn-around times. Sodium hybride generation systems yielding fast and highly accurate results are applied to the analysis of selenium and arsenic. This procedure is so recent that EPA has not yet accepted its application to drinking water evaluations. Often these technological developments and their capabilities are happening so fast that regulatory agencies cannot keep pace with them.

Wet chemistry and microbiology have become highly standardized to the extent that commercially available prepackaged kits and supplies greatly reduce the preparation time and handling steps involved in these analyses. Laboratory support services increasingly rely on disposable materials, thereby reducing labor costs, spurious contamination and analyst error.

Insufficient attention to quality control is a serious hazard when relying on packaged and/or standardized methods in a heavily loaded analytical operation. The importance of quality control cannot be overemphasized because the product of an analytical laboratory is numbers based on test results, which must be accurate to justify the cost of running the facility.

Quality control begins and ends with accurate records. In between, standards (internal and external), split samples, spiked samples, continued staff training, routine and scheduled instrument maintenance, procedure review and methodology updating should be facts of life for all analysts. Quality control means generation of standard curves, determination and performance ranges, evaluation of standards and review of data prior to release. Laboratories that spend less than 20% of their daily effort on quality control are asking for trouble. A properly managed analytical laboratory must place productivity second to accuracy because inaccurate numbers will invariably generate greater problems.

Increasingly, organic compounds are being identified as problematic in groundwater contamination. Skilled chemists using advanced equipment are necessary to properly evaluate contaminant loading. Over the last decade, gas chromatography and column technology (Figure 9) have progressed to the point that organic substances below the parts per million level are routinely detected. Larger laboratories are combining their gas chromatography

Figure 9. Gas chromatography of a sample following initial purge and trap pretreatment.

with mass spectrometers and can scan a broad range of pollutants in a single run. This requires a considerable investment in facilities, equipment and staff training. These factors, along with profit and overhead, must be considered in establishing the laboratory fees for these services, which are frequently substantial. Accurate transcription and timely communication of data are the end results of laboratory participation in a project. A quality assurance review by supervisory personnel to catch errors of transcription and calculation is highly recommended because the data generated must be appropriately reported to the project manager or client. Finally, the analytical process is completed by preparing accurate copies of all data reports and safely storing them for any subsequent review.

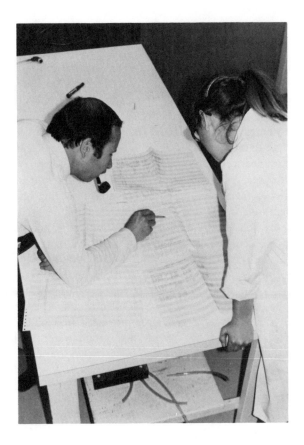

Figure 10. Data review and evaluation following computerized data processing.

Data Interpretation

The final phase of a groundwater monitoring program involves the workup and use of laboratory data and field logs in evaluating conditions at a site (Figure 10). An experienced hydrogeologist will use these data to develop options regarding the nature, extent and treatment of a groundwater system. This review often involves input by regulatory agencies and may result in additional wells, sampling programs, purging and treatment of groundwater, or any number of other actions dictated by circumstances, statute and the project budget.

CONCLUSIONS

Groundwater impairment is a major environmental concern of federal, state and local government. Central to any regulatory efforts are appropriate compliance standards. The equitable application of scientifically sound and technically feasible concentration limits is the only practical way that the groundwater resource can be protected and still maintain a healthy economic climate. The regulatory agencies responsible for accomplishing this goal face several serious concerns.

The groundwater consultant also has a substantial role in developing technical means to meet compliance limits while providing his clients with affordable operational options. Close coordination and negotiation between regulatory agents and the hydrogeological consultant is necessary to solve contamination problems while they are localized and more easily managed. Data generated in the analytical laboratory support both the regulatory effort and the problem-solving activities of the consultant. Assuming sound sample collection and appropriate project conceptualization, the laboratory data serve as the barometer of the project's effectiveness. These components, together with a client who is genuinely interested in avoiding or correcting groundwater problems, can assure that groundwater quality will be maintained or restored.

REFERENCES

1. "Michigan Water Resources Commission Act," *Public Acts of the Legislature* 245:597-599 (1929).
2. "Safe Drinking Water Act," *Public Acts of the Legislature* 399: 325.1001-325.1023 (1976).
3. "Safe Drinking Water Act," USC PL 93-523, Government Printing Office, Washington, DC (1979).
4. "Environmental Protection Agency National Interim Drinking Water Regulations," 40 *Federal Register* 141; 40 FR 59565 (December 24, 1975); Amended by 41 FR 28402 (July 9, 1976); 44 FR 68641 (November 29, 1979), U. S. Government Printing Office, Washington, DC (1980).
5. Michigan Department of Natural Resources, Resource Recovery Division. "Maximum Permissible Concentrations and Recommended Limits in Drinking Water Standards," Unpublished results (1979).
6. Pipes, W. O., Ed. *Water Quality and Health Significance of Bacterial Indicators of Pollution,* Proceedings of a Workshop held at Drexel University, Philadelphia, PA, April 17, 18, 1978 (Philadelphia: Drexel University, 1978).

7. Demopoulous, H. "Environmentally Induced Cancer . . . Separating Truth from Myth," speech to the Synthetic Organic Chemical Manufacturers Association, Inc., Hasbrough, NJ, October 4, 1979.
8. Higginson, J. "Cancer and Environment: Higginson Speaks Out," *Science* 205:1363-1366 (1979).
9. Michigan Department of Natural Resources, Water Resources Commission. "Part 22. Ground Water Quality Rules (Proposed)," Lansing, MI (1978).
10. Michigan Department of Natural Resources, Environmental Services Division. "Critical Materials Register 1978," Lansing, MI (1978).
11. *Standard Methods for the Examination of Water and Wastewater,* 14th ed. (New York: American Public Health Association, 1976).
12. "Methods for Chemical Analysis of Water and Wastes," EPA-600/4-79-020, Government Printing Office, Washington, DC (1979).
13. *1979 Annual Book of ASTM Standards, Part 31, Water* (Philadelphia: American Society for Testing and Materials, 1979).

CHAPTER 20

LAND DISPOSAL EFFECTS ON GROUNDWATER

William A. Kelley, PE, Chief
   Division of Water Supply
   Bureau of Environmental and Occupational Health
   Michigan Department of Public Health
   Lansing, Michigan

The disposal of sludge as a waste versus the utilization of sludge as a resource—many communities and industries must make a conscious decision about which option to pursue. Whichever is chosen, proper control must be exercised to avoid possible groundwater contamination. The contamination of Michigan's groundwater is a subject of current concern across the state because municipal and industrial wastes have degraded the groundwater in various areas. Industrial chemicals and/or wastes have proved to be particularly critical. When waste organic or inorganic chemicals enter an aquifer, people using the aquifer for their drinking water may as well look for another source of water because of the extended period of time needed to cleanse the contaminated aquifer. This emphasizes the need to exercise great care in protecting our groundwater resource.

Groundwater contamination problems have developed due to spills of chemicals, either accidental or routine; inadequate storage; and waste disposal into or onto the ground. In some areas, these spills and disposal problems have caused degradation of what have been usable aquifers.

DEVELOPMENT OF GROUNDWATER QUALITY STANDARDS

The Department of Natural Resources has developed proposed "Ground Water Quality Standards," which would adopt the philosophy of nondegradation. The need for development of Ground Water Quality Standards gained impetus with the discussion of revised Water Quality Standards for surface waters. The proposed revisions to Michigan's Water Quality Standards were so stringent that they likely would have forced increased land disposal of wastes, and many people were concerned that the increased utilization of land disposal for waste products would create more widespread problems with groundwater quality across the state. The philosophy of nondegradation, which is delineated in the proposed Ground Water Quality Standards, would be flexible in that the proposed standards do not go to the point of requiring total nondegradation. The Department of Natural Resources can allow a variance; however, issuing a variance calls for a responsible, professional judgment. It would be essential for the subject project to be reviewed in detail in advance, and this would make it possible for people to know a particular disposal project might adversely affect groundwater quality. This is much preferable to determining, after the fact, that wells have been degraded unknowingly.

It has been stated that the present groundwater contamination problems around the state are a result of a historical lack of knowledge or concern. However, if there was any lack of concern, it most likely developed due to a lack of knowledge about groundwater and the fact that it is vulnerable to contamination. The important thing is to learn from past mistakes. There is certainly no one answer for all types of wastes or all types of disposal sites. Some groundwater formations are tight soils, and heavy metals will not migrate to a groundwater formation. In other areas, heavy metals will penetrate into a usable groundwater aquifer rapidly and can be expected to cause problems for those people who might be utilizing a groundwater supply immediately downstream. We simply cannot afford to take unnecessary chances in areas where groundwater is unprotected and is usable in quality; site selection is important and should be the first item to be propsed, reviewed and either approved or disapproved. Design can proceed after the site is determined to be acceptable.

SITE SELECTION

It is essential to have good hydrogeological studies completed, reviewed and evaluated before considering approval of a site for waste disposal on the land. Monitoring wells for a particular disposal site must be designed, properly installed and routinely sampled if they are to be of any value at all. The monitoring must be initiated before the site is used for waste disposal to develop meaningful background data. If monitoring wells are simply holes in the ground installed as a required part of a waste disposal project but not used on a routine basis, they are a waste of money and will provide no benefit to the public who will be paying for them.

As a precaution, Michigan's Department of Public Health is working with its Department of Natural Resources to determine where municipal sludge is being used as a soil conditioner on farm land. This information will be used during the evaluation process for potential well sites for communities and other public supplies. An effort is made to keep large-capacity production wells away from all disposal sites where potential contaminants could be located. The Department of Natural Resources is continually being advised of the location of public well supplies, including new well installations. Monitoring should be required for even this type of disposal/utilization if there is any question as to the capability of the soil at a particular site to adequately tie-up chemical contaminants that may be found in the sludge.

Landfills are discussed routinely as possible sludge disposal sites. Public health agencies are concerned when landfills are located on or adjacent to usable aquifers. It is likely that public health agencies would consider those aquifers unacceptable or no longer usable if there is any possibility of hazardous wastes being disposed of in the landfill. Dependence on a liner (clay or other) is unacceptable when considering the absolute protection of the groundwater quality. Site location is *critical!* It is essential that disposal sites *not* be located on sand and gravel aquifers or other vulnerable aquifers with the attempt made to protect the aquifers by trying to seal the disposal sites with clay, etc. It is strongly recommended that site selection be handled at the outset by locating areas where there is a natural clay barrier or where the groundwater is of such poor quality that it is not used for drinking and household uses. Then the necessary steps should be taken to seal the potential site and use it for disposal. As mentioned earlier, we must learn from past mistakes.

CLASSIFICATION OF WASTES

Some federal and state legislation and regulations classify wastes in part by reference to the U. S. Environmental Protection Agency's (EPA) Interim Primary Drinking Water Standards. This is a less than satisfactory concept because (1) the present drinking water standards are not all-inclusive and do not cover a great number of parameters; and (2) the contention that wastes will not be classified as hazardous until the concentration of contaminants is ten times the drinking water standards tends to indicate that it is accept-

able to degrade the groundwater—at least up to the drinking water standards—
a philosophy that is *not* acceptable from the public health perspective.

SUMMARY

Some regulations are necessary as our society becomes more complex.
Our earlier discussion of the Water Quality Standards and how those proposed
standards brought about the discussion of the Ground Water Quality Standards indicates how regulations (particularly stringent, narrow point-of-view
regulations) can breed more regulations. The public water supply program
in Michigan managed to do an efficient job of protecting public health for
many years with few administrative rules. While it may be wishful thinking
to feel we could continue to operate in that fashion, it certainly was preferable in many ways to the threat of overregulation now facing us in the pres-
ent-day world. Great care should be taken to ensure that regulations are
neither requested nor approved unless there is a documented need for them.
This is a difficult task when dealing with federal environmental programs
because the impact of the federal bureaucracy is tremendous.

The debate as to whether wastewater sludge is a resource to be utilized
or a waste to be disposed of properly will undoubtedly continue. It is possible
to summarize the discussion simply by saying that if sludge can be used as
a resource *and* properly controlled to avoid contamination of water, crops,
animals, it is an excellent method of disposal.

CHAPTER 21

CADMIUM EFFECTS ON MYCORRHIAGE: A RESEARCH NEED IN FORESTLAND APPLICATION OF SLUDGE

Jonathan W. Bulkley

Professor of Civil Engineering and Natural Resources
The University of Michigan
Ann Arbor, Michigan

Since the passage of The Water Pollution Control Act Amendments of 1972 (PL 92-500) and the Clean Water Act of 1977 (PL 95-217), municipalities throughout the United States have been working to improve the performance and capability of their wastewater treatment plants. One goal of the federal legislation has been for all municipalities to provide a level of treatment at least equal to secondary treatment. One direct result of the enhanced treatment of wastewater is a significant increase in the production of solids (sludge) at these treatment plants. Consequently, one component of improved performance of the treatment plants here and elsewhere in the world is to implement means of reliable disposal of the sludge removed from the wastewater flow at the plant.

For example, the city of Detroit operates a major wastewater treatment plant that produces an average of 650-700 dry tons of solids per day. During Detroit's Step I Segmented Facility Plan process, land application was considered for sludge disposal. While that particular alternative was not chosen for implementation, it does indicate the importance of evaluating the application of sludge to land. In particular, the application of sludge to forest land appears attractive because, at first examination, it may minimize the chance for bioaccumulation of toxic heavy metals in human food chains.

263

Issues associated with land application of municipal wastewater treatment plant effluents and sludges to forest land are important research topics. This chapter represents a brief status report on one component of this important issue, namely, the fate of heavy metals that may be present in the wastewater treatment plant sludge. In particular, it is important to establish whether the application of these sludges to forest land will minimize the possible adverse environmental impacts of toxic heavy metals. It is important to establish under what conditions forest land application of sludge containing toxic heavy metals will be effective and environmentally sound.

The urgent need to examine land application of municipal sludges to ensure that one is simply not transferring an environmental hazard from one location to another has been clearly stated by Moore in his introduction to this book. It is particularly important in land application of sludge to take appropriate steps to ensure that toxic heavy metals or other pollutants are controlled so that they do not enter the human food chain. Accordingly, this chapter will focus on research findings regarding the observed distribution of heavy metals in forest land.

It is appropriate to note that sludge disposal is a world wide problem. Table I provides a report on current sludge disposal practice in the United Kingdom. It is important to note that 68% (on a dry ton basis) of Britain's annual sludge production is handled by land disposal. Also, because of the high costs of energy, only 4% of the sludge is incinerated. With increasing energy costs and a ban on ocean disposal of sludge, environmentally sound

Table I. Sludge Disposal in the United Kingdom [1]

Annual Total Sludge Production (dry tons - solids), 1.25×10^6 ton/yr

Land Disposal	Incineration		Sea
68%	4%		28%

	Land Disposal (utilization), 46%		
Grazing	Arable	Reclamation	Minor Uses
15%	24%	4%	3%

Land Disposal (Nonproductive Disposal)
Landfill
22%

land application of wastewater treatment plant sludge may become the long-term disposal method in many more areas of this country. It is clear that the federal, as well as many state, governments are actively encouraging land application of both sludge and effluents. Nevertheless, one must proceed with care. The final decisions regarding long-term disposal to forest land of sludge containing heavy metals must be based on a careful analysis of the facts regarding the ability of the disposal process to bind and isolate the heavy metals for extended time periods.

CURRENT DEVELOPMENTS

Overview

Sopper and Kerr [2] have reported on forest land applications of sludge in a number of areas in the United States. In addition, Edmonds and Cole [3] have provided a comprehensive statement on the current status of a University of Washington research effort underway since 1973. This study is investigating the feasibility of applying dewatered sewage sludge to a forest soil. Both provide an excellent source of important information on the current state-of-the-art with regard to sludge utilization in forest land in the United States. Evidence presented in these two sources indicates that trees do respond in a significant manner in terms of enhanced growth as a result of sludge application. The forest represents a renewable resource that can be utilized for the benefit of society, and the enhanced production of wood and timber as a result of sludge application is certainly encouraging.

Heavy Metals

Background

While much research has been directed toward overall response of the forest to sludge application, there are a number of questions that need further exploration. In particular, it is important to understand the natural processes governing the circulation of heavy metals and other toxics that could be introduced into the forest ecosystem as a result of sludge application.

In 1976, Sidle and Sopper [4] reported that there was limited information on the fate of cadmium in forest ecosystems. Their study was directed toward determining whether cadmium levels were increasing at sites treated with wastewater effluent and sludge during the decade of 1964-74. Their results indicated that the cadmium concentration in trees did not differ significantly between the treated area and the control area. However, the cadmium concentration in trees did vary according to tree species. Also, the researchers reported that the cadmium concentration in the 0- to 5-cm depth of soil from the treated area had a significantly higher cadmium concentration than the same soil profile in the control area.

Munshower [5] reported on a comparison of two sites. One site was located 24 km downwind from a smelter complex, which had operated for 75 years. The second site had no history of industrial contamination. In this specific case, the cadmium application to the soil came as a result of air pollution from the smelter complex. Basically, the research results demonstrated that soils, plants and animals from the polluted study area all had significantly higher cadmium concentrations compared to the nonpolluted site. The investigator noted that even after the smelter had been closed (and the source of the cadmium contamination removed), levels of cadmium remained significantly elevated in plants and animals in the polluted site. One explanation of this latter observation is that the soil in the polluted site served as a reservoir for cadmium that had accumulated over the operation period of the smelter.

Impact on Trees

The research cited demonstrates the following:

1. Sludge application to forest land enhances tree growth.
2. Cadmium applied to land accumulates in soil, plants and animals.
3. The cadmium appears to concentrate differentially in trees, i.e., different species appear to accumulate more cadmium than others.
4. Cadmium is toxic to tree species as well as to plants.

Lamoreaux and Chancy [6] demonstrated that increasing levels of cadmium had very adverse impacts on silver maple seedlings. In particular, the dry weight of the trees decreased with increasing cadmium applications. Also they reported a significant reduction in height. The trees showed signs of wilting even though they were well watered. The increased cadmium levels reduced the water-conducting capacity of the silver maple stems. Accordingly, as the application levels of cadmium increase, toxic effects are observed in trees.

The research of Kelly et al. [7] confirmed the observations of Lamoreaux and Chaney. This effort did note higher cadmium concentrations in the root and shoot tissues of all evaluated species. The elevated concentration came as a result of increased cadmium concentration in the soil. At the highest levels of cadmium application, reduction in tree height was observed.

Finally, all species showed a decrease in root biomass with increasing cadmium concentration. Kelly goes on to observe that the reduction in root biomass may have adverse implications over the long term for aboveground production and the survival rate of seedlings. Also, since the cadmium in Kelly's study appeared to be relatively immobile in the soil, the research finding raises the important question of long-term forest productivity if heavy metals are being introduced into the forest ecosystem through sludge application. Kelly's study also raised the question of the potential hazard that may exist as a result of animals feeding on contaminated plant tissue.

Ecosystem Function

Van Hook et al. [8] examined the distribution and cycling of heavy metals in a forest. If one recognizes that ecosystems recycle essential elements, then, as Van Hook et al. point out, the very same complex of biological processes responsible for effective ecosystem function also influence the transport and accumulation of potentially toxic heavy metals. Their study found that 28% of the cadmium applied to the forest concentrated in 11% of the biomass of the forest. Further, this significant concentration is in the critical root system. Previous work has provided evidence of the toxic impact of heavy metals on plants and trees. The Van Hook study offers evidence that these toxic heavy metals accumulate at critical pathways (such as roots) and may impair the function of the ecosystem. Part of this impairment may result from the adverse influence the toxics have on the mechanisms of nutrient transport.

For example, one critical relationship that may be hindered by heavy metal accumulation occurs at the soil–root interface. This is the symbiotic relationship that exists between tree feeder roots and mycorrhizae. Spurr and Barnes [9] provide an excellent description of the critical function of mycorrhizae in assuring successful growth of many tree species. Mycorrhizae exist at the soil–root interface and serve to accumulate nutrients and water for transmission to the tree through the root system. It should be noted that heavy metals such as cadmium are often used as fungicides. Accordingly, one must examine whether long-term application to forest land of sludge containing heavy metals toxic to fungi may, in fact, result in the accumulation of sufficient heavy metals at the soil–root interface to reduce the mycosymbiont activity between the mycorrhiza and the tree feeder roots. This issue has been identified by Harris [10]. If the mycosymbiont activity is reduced or eliminated, then the impact on tree growth could be extreme. For example, if the trees did not survive, the heavy metals presently bound within the soil and especially within the root zone might eventually be released to the environment as a result of erosion or seepage to groundwater.

Svoboda et al. [11] confirm the accumulation of cadmium in roots of trees from sludge applications. This research also indicates that certain tree species may be more desirable than others where forest land is to receive significant sludge application. For example, their findings indicate that silver maple, eastern white pine and green ash are particularly attractive because each of these species has the following characteristics:

1. Each tends to accumulate relatively large amounts of heavy metals in the roots.
2. Each appears to have reasonable growth rates in the presence of the heavy metals.
3. The accumulation of metals in the foliage of each was usually low compared with other species tested.

This latter attribute would minimize the bioaccumulation of cadmium in animals feeding on the tree foliage. However, these researchers clearly state that ecosystem processes may be severely altered over long periods of time as a result of heavy metals accumulation. Careful evaluation over a longer time period is required to ensure that the heavy metals do not become toxic to essential ecosystem processes.

Sundberg et al. [12] reported on the changes in microfungal populations in soil following application of sludge to strip mine spoils. This situation is special because the strip mine spoil itself is nearly void of microfungal populations from the acid conditions of the soil. This research indicates a tenfold increase in the population density of microfungal populations following application of wastewater treatment plant sludge to the strip mine spoil. However, these researchers also observe that since different fungi have different tolerances to heavy metal concentrations, heavy metal toxicity may be very important in controlling species or racial composition of the fungal population.

OBSERVATIONS AND SUMMARY

As a direct result of improved treatment of wastewater, municipalities have increased the quantity of sludge that needs to be disposed of in an economic and environmentally sound manner. Municipal sludges may contain toxic heavy metals as well as other pollutants. It is prudent engineering to consider disposal means for these sludges so as to minimize the release of these heavy metals into the human and animal food chain.

Land application of sludge containing heavy metals for the purpose of growing trees appears to be an attractive alternative for long-term sludge disposal. First, this alternative involves the enhanced production of trees as opposed to crops for either human or animal consumption. Accordingly, this disposal method should minimize the food-chain accumulation of heavy metals. Secondly, certain tree species appear to be tolerant of heavy metals and concentrate the heavy metals in the root biomass. As long as one maintains a viable tree system, the heavy metals should remain bound and not available to contaminate either groundwaters or surface waters.

However, to assume long-term viability of the forest treated with sludge containing heavy metals it may be necessary to adopt conservative sludge application rates to minimize toxic impacts on the mycorrhizae. It may be desirable and necessary to monitor fungal populations as a control factor for sludge application. Critical changes in the mycorrhizae would occur before the impact would be observed on the trees themselves. Furthermore, long-term research and monitoring of sites should be directed toward increasing our understanding of high levels of heavy metals availability and the subsequent impact of these metals on necessary mycosymbiont activity.

If we fail to properly examine the relationship between the vital ecosystem functions contributing to forest growth and the application of toxic heavy metals to forest lands, then we may fail to understand the limitation and constraints that must be associated with the proper disposal of sludges containing these toxic heavy metals. Consequently, a risk will be incurred. This risk is simply that the forest growth will be severely limited because of accumulated toxic effects from the heavy metals. Further, if the trees fail to live, it is conceivable that the heavy metals that had been bound in the soil–root system will be released into the environment as a result of erosion or leaching activity.

Finally, the United States Environmental Protection Agency (EPA) [13] indicate that one concern regarding the future of land treatment relates to our ability to define the capacity of various treatment systems to control trace organics and toxic elements, i.e., heavy metals. EPA's present perspective is that soil treatment systems will play an important role in this area. This brief chapter indicates that it will be important to understand and monitor critical ecosystem functions involving forest land applications if this approach is to have long-term viability and effectiveness.

REFERENCES

1. National Water Council. "Sludge," *Water (Suppl.)* (November 1979).
2. Sopper, W. E., and S. N. Kerr, Eds. *Utilization of Municipal Sewage Effluent and Sludge on Forest and Disturbed Land* (University Park, PA: The Pennsylvania State University Press, 1979).

3. Edmonds, R. L., and D. W. Cole, Eds. "Use of Dewatered Sludge as an Amendment for Forest Growth: Management and Biological Assessments," *Bulletin No. 3,* Center for Ecosystem Studies, College of Forest Resources, University of Washington, Seattle, WA.

4. Sidle, R. C., and W. E. Sopper. "Cadmium Distribution in Forest Ecosystems Irrigated with Treated Municipal Wastewater and Sludge," *J. Environ. Qual.* 5(4):419-422 (1976).

5. Munshower, F. F. "Cadmium Accumulation in Plants and Animals of Polluted and Non-Polluted Grassland," *J. Environ. Qual.* 6(4):411 (1977).

6. Lamoreaux, R. J., and W. R. Chaney. "Growth and Water Movement in Silver Maple Seedlings Affected by Cadmium," *J. Environ. Qual.* 6(2):201-202 (1977).

7. Kelly, J. M., G. R. Parker and W. W. McFee. "Heavy Metal Accumulation and Growth of Seedlings of Five Forest Species as Influenced by Soil Cadmium Level," *J. Environ. Qual.* 8(3):361-362 (1979).

8. Van Hook, R. I., W. F. Harris and G. S. Henderson. "Cadmium, Lead, and Zinc Distribution and Cycling in a Mixed Deciduous Forest," *Ambio* 6(5):281-286 (1977).

9. Spurr, S. H., and B. V. Barnes. *Forest Ecology,* 3rd ed. (New York: John Wiley & Sons, Inc., 1980), pp. 221-225.

10. Harris, W. F. Personal communication (July 10, 1979).

11. Svoboda, D., G. Smout, G. T. Weaver and P. L. Roth. In: *Utilization of Municipal Sewage Effluent and Sludge on Forest and Disturbed Land,* W. E. Sopper and S. D. Kerr, Eds. (University Park, PA: The Pennsylvania State University Press, 1979), pp. 395-405.

12. Sundberg, W. J., D. L. Borders and G. L. Albright. In: *Utilization of Municipal Sewage Effluent and Sludge on Forest and Disturbed Land,* W. E. Sopper and S. N. Kerr, Eds. (University Park, PA: The Pennsylvania State University Press, 1979), pp. 463-469.

13. United States Environmental Protection Agency, "A History of Land Application as a Treatment Alternative," MCD-40, EPA 430/9-79-012, Washington, DC (1979), p. 53.

SECTION V

REGULATORY ASPECTS

CHAPTER 22

FEDERAL REGULATORY CONSIDERATIONS IN SLUDGE UTILIZATION

Gregory A. Vanderlaan

Municipal Sludge Coordinator
Water Division
U.S. Environmental Protection Agency
Chicago, Illinois

We can all agree that there are no easy answers to the safe deposition of certain materials in our environment. Further, we all recognize that we have a multifaceted issue on our hands whose resolution can only be achieved through the holistic perspective advocated by the early environmentalists. Sludge management is directly related to such other societal growth problems as the availability of land and diminishing supplies of materials and energy. Since 1970 our fastest population growth has been in rural areas, not in the cities. People are moving out of our largest cities in substantial numbers into counties with the lowest population densities. Rural planning capabilities and land use controls, where they exist, cannot anticipate or regulate the rapid pace of new development. Further, rural fiscal resources often cannot meet the demands for such public services as sewers, schools, medical facilities, police and fire protection, which are created by new residents.

Obviously, the many tasks confronting municipalities experiencing population increases include planning sewage treatment improvements—expansions or entirely new systems. One of the first tasks in the planning stage should be the development of a sludge management program. The end product of the wastewater treatment operation should be dealt with at the very beginning of the planning process. Municipal sludge treatment and handling should be an issue of public decision-making and should be brought out of the conference rooms of sanitary engineers and contractors who have been trained in designing and building bigger and better treatment systems. In these communities there have to be citizens trading ideas about what kind of future sludge management programs are best for their area.

APPROACH

Decisions must be made now in many communities for the purpose of selecting the most suitable techniques for sludge management. To assist communities in meeting this challenge the U.S. Environmental Protection Agency (EPA) formulated a policy (1) to achieve the implementation of acceptable and economical sludge management options so as to protect public health, welfare and the environment; and (2) to conserve natural resources through beneficial utilization of sludges.

There are four problem areas that communities must resolve prior to selecting a sludge management scheme. These are:

- public health issues
- technological factors
- intermedia issues
- social/economic/institutional

EPA guidance developed through administration of the Clean Water Act Construction Grants Program and regulations promulgated under both the Clean Water Act and the Resource Conservation and Recovery Act are designed to assist communities in their efforts to resolve these issues.

Recognizing that land application of municipal sludge is often less costly than other handling alternatives and results in agricultural benefits from its fertilizer value and soil conditioning value, Congress has encouraged and mandated sludge recycling in the Water Pollution Control Act Amendments of 1972 and the Clean Water Act of 1977. EPA construction grant regulations specifically require municipalities planning new treatment systems to evaluate "land application techniques." Furthermore, under the innovative and alternative technology guidelines of these regulations, EPA provides financial incentives for implementation of sludge land application programs by designating this technique an alternative process. Sludge handling systems and supplemental processing required to implement the program are eligible for federal funds at the 85% level. Sludge handling systems and supplemental processing include processes to significantly reduce pathogens, i.e., anaerobic digestion or composting. Under certain conditions and subject to certification by the state agencies, EPA will fund the purchase of land when an area has been dedicated for sludge processing.

Municipalities developing systems with grant funds need to indicate clearly in their plan of operation that the sludge management will comply with appropriate federal regulations for land application. The specific federal requirements are found under 40 CFR Part 257 entitled "Criteria for Classification of Solid Waste Disposal Facilities and Practices." Many of these requirements are based on EPA's sludge technical bulletin entitled "Municipal Sludge Management: Environmental Factors." This bulletin addresses important factors relevant to the various sludge management options for land utilization. The management approaches put forward were designed to allow the use of sludge from municipal wastewater treatment plants with minimal health risk. To ensure this, the technical bulletin recommended the following:

1. Soil pH should be maintained at 6.5 or above where sludge is applied.
2. Sludge stabilization should be equivalent to anaerobic digestion for 10 days at 95°F.
3. Sludge should be applied based on the crop's nitrogen need to protect groundwater.
4. Rates of sludge application should be consistent with the recommended heavy metal annual and total accumulations criteria.

The sludge bulletin identified three categories of agricultural use for sewage sludge, with various levels of monitoring recommended for the specific category. These categories are:

1. privately owned agricultural lands,
2. publicly controlled agricultural lands, and
3. sites dedicated for disposal

There were two major reasons for developing these categories for sludge use: (1) to ensure maximum utilization for a wide range of sludge quality, and (2) to suggest a level of monitoring commensurate with sludge application and heavy metal concentration. For example, it was recommended that sludge application rates be low on private lands and higher on dedicated sites. The intent was to increase the level of control as the potential for pollution becomes greater.

REGULATIONS

The technical sludge bulletin required five years to develop. These efforts were relied on extensively in the development of "Criteria for Solid Waste Disposal Facilities," forming the basis of control for disposal of all nonhazardous solid waste on land, including municipal sludge. The land application portion of the "Criteria" was published in interim final form in the *Federal Register* on September 13, 1979 under authority of Section 4004 of the Resource Conservation and Recovery Act of 1976 and Section 405(d) of the Clean Water Act of 1977. Compliance with the "Criteria" is mandatory under Section 405(e) of the Clean Water Act. The "Criteria" directly involve sludge quality in three areas: cadmium loadings, polychlorinated biphenyls (PCB) concentrations and pathogen levels.

Cadmium

The "Criteria" put forth two approaches for contolling cadmium levels. The first approach involves application site management controls and standards governing cadmium applications. It requires that the soil/sludge mixture pH be 6.5 or greater at the time of each sludge application. There are no pH requirements if the sludge contains concentrations of cadmium of 2 mg/kg (dry weight) or less. Where the application of sludge to soils will be used for the production of tobacco, leafy vegetables or root crops grown for human consumption, and annual loading limit of 0.5 kg/ha may not be exceeded. For all other food crops, the annual cadmium application rate may not exceed the rate shown in Table I.

Table I. Annual Cadmium Limits for Food-Chain Crops

Time	Annual Cadmium Application (kg/ha)
Present- June 30, 1984	2.0
July 1, 1984- December 31, 1986	1.25
Beginning January 1, 1987	0.5

Limitations on the total cumulative sludge cadmium applications are determined by the soil pH and the soil cation exchange capacity (CEC). There are three soil categories based on pH: (1) those with natural pH of 6.5, or above; (2) those with natural pH below 6.5; and (3) those with natural pH below 6.5 but where pH will be maintained at or above 6.5 for as long as food-chain crops are grown. Soil CEC is an easily measured index of those properties (based on the nature and content of clay and organic matter) that affect the soil's ability to adsorb cadmium. High CEC levels indicate that a soil has a greater capacity to adsorb cadmium, thus preventing the cadmium from entering plants grown on sludge-amended soil. Evidence available to EPA indicates that the CEC is a particularly important factor in limiting cadmium uptakes in high pH soils. In highly acidic soils, however, pH becomes the dominant factor. Based on these factors, cumulative loadings are as shown in Table II.

Table II. Maximum Cumulative Loading Limitations (kg/ha)

Soil Cation Exchange Capacity (meq/100 g)	Background Soil	
	pH<6.5	pH≥6.5
<5	5	5
5-15	5	10
>15	5	20

In cases in which the background soil pH is less than 6.5 but is adjusted and maintained at 6.5 or above when the food-chain crop is grown, higher cumulative loading limits are feasible as shown in Table III.

Table III. Maximum Cumulative Loading Limitations (kg/ha)

Soil Cation Exchange Capacity (meq/100 g)	Background Soil pH<6.5 but Adjusted to≥6.5
<5	5
5-15	10
>15	20

The second approach for controlling cadmium levels is known as the "controlled site" or "dedicated site." This concept is utilized at the Fulton County project under the direction of the Metropolitan Sanitary District of Greater Chicago. With this approach, the cadmium application rate is unrestricted, provided the following four requirements are met:

1. Only animal feed may be grown.
2. The slude and soil mixture must have a pH of 6.5 or greater at the time

of sludge application or at the time the crop is planted, whichever occurs later.

3. There must be developed a facility operating plan demonstrating how the animal feed will be distributed and what safeguards will be utilized to prevent the crop from becoming a direct human food source.

4. There must be a stipulation in the land record or property deed stating that the property has received sludge at high cadmium application rates.

Polychlorinated Biphenyls (PCB)

For municipal sludges containing PCB in concentrations greater than 10 mg/kg (dry weight), the "Criteria" requires that such sludge be incorporated into the soil. "Incorporated into the soil" is defined as injection or mixing of the sludge beneath the surface of the soil. Incorporation into the soil is not required if assurance can be given that the PCB content in the animal feed grown is less than 0.2 mg/kg or that milk from animals grazed on land that has been amended with sludge that has less than 1.5 mg/kg of PCB. These concentrations are based on tolerance levels established by the Food and Drug Administration (FDA) in defining health risks.

Pathogen Levels

Sludge contains pathogenic bacteria, viruses and parasites, which can infect both humans and animals. There are two pathogen reduction practices required for different land application techniques:

1. Septage (solids from septic tanks) may be applied directly to agricultural lands provided that public access is restricted for 12 months and that the grazing by animals whose products are consumed by humans is restricted for at least a month. Similarly, sewage sludge that has achieved a level of pathogen reduction comparable to anaerobic digestion may be applied directly to agricultural land provided that public access is controlled for at least 12 months and grazing by animals whose products are consumed by humans is prevented for at least a month.

2. If sewage sludge or septage is applied to land used for crops directly consumed by humans, a stabilization process equivalent to heat-drying or thermophilic anaerobic digestion must be used. This level of treatment is not required if there is no contact between the waste sludge and the edible portion of the crop. In this case, however, the sludge is to be treated to the level as indicated above.

Other Criteria

It is important to remember that all portions of the "Criteria" found within 40 CFR 257 apply to land application programs. Additional portions of the "Criteria" describe performance standards and/or operating techniques to protect air and water quality, sensitive lands or biological resources, and public health and safety. In floodplains, for example, the "Criteria" would prohibit land application of sludge, which would restrict flow of the base 100-year floodplain, reduce temporary water storage capacity or result in a washout of sludge that would threaten human life, wildlife, or land or water resources.

FUTURE REGULATIONS FOR SLUDGE GIVEAWAY/SALE

New regulations are currently being developed governing other uses of municipal sewage sludge. These new regulations will be issued under the authority of Section 405(d) of the Clean Water Act and are expected to address the following:

- General monitoring requirements
- Landspreading for reclamation
- Landspreading to food- and nonfood-chain crops
- Giveaway/sale/home use practices
- Landfilling practices
- Disposal in surface impoundments
- Thermal conversion (incineration/heat treatment, etc.)

It is anticipated that these regulations will identify a "good sludge" for widespread use with no restrictions based on minimal requirements. Minimal requirements would include limiting concentrations of cadmium, PCB, lead and zinc. Scientific support for this identification will be needed from agronomic studies.

ECONOMICS

Currently in the U.S., 31% of our treated sludge is applied to the land. If by 1990, we were to apply 50-75% of the treated sludge generated, the estimated savings that could be realized from this increase in land utilization would range from $100-$500 million annually (1978 dollars). Most of the savings is estimated to come from the nutrient value of the sludge itself.

Approximately $93 million worth of nitrogen, phosphorus and potassium as fertilizer were present in sludges in 1978. Energy is required to manufacture fertilizer that is otherwise used in place of sludge. To produce the nitrogen destroyed by sludge incineration would require 420 gallons of crude oil per ton, or 23 million gallons of oil to manufacture 55,000 tons of nitrogen wasted in 1978. The estimated consumption of No. 2 fuel oil to burn one ton of dry sludge was found to be 50 gallons, or about 120 million gallons to burn the 2.4 million dry tons incinerated in 1978. The total energy involved was about 143 million gallons, enough oil to heat 140,000 homes in Minneapolis.

SUMMARY

Land application of sludge is consistent not only with the philosophy of returning these materials to natural cycles from which they were generated, but also with the economic realities of energy use and transfer. Clearly, it makes sense both environmentally and economically to land apply municipal sewage sludge.

REFERENCES

1. Ehreth, D. J. "Municipal Sludge Management: Problems and R & D," paper presented at the Conference on Evaluation of Current Developments in Municipal Waste Treatment, Baltimore, MD, January 26-27, 1977.
2. The Council on Environmental Quality. "Environmental Quality—The Ninth Annual Report of the Council on Environmental Quality," Washington, DC (1978).
3. Goldstein, J. *Sensible Sluge—A New Look at a Wasted Natural Resource* (Emmaus, PA, Rodale Press, 1977).
4. "Municipal Wastewater Treatment Works—Construction Grants Program 40 CFR Part 35, *Federal Register* (September 27, 1978).
5. U.S. Environmental Protection Agency. "Municipal Sludge Management: Environmental Factors," EPA 430/9-77-004; MCD-28 (1977).
6. Walker, J. M. "Government Regulations on the Use of Municipal Organic Materials on Agricultural Lands," *Hort Science* (In press).
7. Walker, J. M. "Using Municipal Sewage Sludge on Land Makes Sense," *Compost Sci. Land Util.* 20 (5) (1979).

INDEX